WOMEN, PEACE AND SECURITY IN MILITARY OPERATIONS

SANDHURST TRENDS IN INTERNATIONAL CONFLICT

Edited by David Brown, Donette Murray, Malte Riemann, Norma Rossi and Max Thompson, The Royal Military Academy Sandhurst, UK

The Sandhurst Trends in International Conflict series is a cutting-edge forum and platform for original thought and debate on military and security matters within the contemporary international security environment. It aims to stimulate authors to think critically about contemporary conflict and security more generally and to identify and evaluate practical, political and doctrinal lessons from recent experience. The Sandhurst series invites practitioners and academics from the fields of security, diplomacy, the law, politics and the military to interrogate and publish on the key debates that will shape both the contemporary international security environment and a modern military operating within it.

Other books in this series

Violent Non-State Actors in Modern Conflict
978 1 912440 20 7

War Amongst the People: Critical Assessments
978 1 912440 02 3

Fragile and Failing States: Challenges and Responses
978 1 912440 19 1

Women, Peace and Security in Military Operations

Edited by
ANDREE-ANNE MELANCON
MAX THOMPSON

Howgate Publishing Limited

First published in 2023 by
Howgate Publishing Limited
Station House
50 North Street
Havant
Hampshire
PO9 1QU
Email: info@howgatepublishing.com
Web: www.howgatepublishing.com

British Library Cataloguing-in-Publication Data
A catalogue record for this book is available from the British Library

ISBN 978-1-912440-43-6 (hardback)

The views expressed in this book are those of the individual authors and do not necessarily reflect official policy or position.

CONTENTS

TABLES

FIGURES

ABOUT THE EDITORS

Dr Max Thompson is Deputy Head of Department in Defence and International Affairs at the Royal Military Academy Sandhurst. He is responsible for oversight of the Army Higher Education Programme, which has included integrating WPS into academic exercises. He completed his DPhil in International Relations at St Edmund Hall, Oxford writing on the role of friendship in US foreign policy and alliance formation. Before arriving at RMAS, he held lectureships at several Oxford colleges and worked as an international consultant at the International Trade Centre in Geneva. He has delivered defence engagement courses on international conflict management, human security in military operations, and counterterrorism all around the world. His research interests centre on alliance theory, US foreign policy, the development of the liberal world order, and human security and the women, peace and security agenda.

Dr Andree-Anne Melancon is a senior lecturer in Defence and International Affairs at the Royal Military Academy Sandhurst. She is responsible for the delivery of law of armed conflict training as well as academic support on field training exercises. She was heavily involved in the inclusion of the WPS agenda into military training. She completed her PhD in Politics at the University of Sheffield where she explored target selection in US drone strikes. Beyond teaching, she is highly engaged in a range of defence engagement activities, having delivered courses on a range of topics to senior officers in over a dozen countries and supported international LOAC competitions. Her research interests include military ethics, the law of armed conflict, human security, and the WPS agenda.

CONTRIBUTORS

Sanam Naraghi Anderlini is the Founder and CEO of the International Civil Society Action Network (ICAN). Born in Iran and educated in the UK, with an MPhil in Social Anthropology from Cambridge University, she teaches at Columbia University's School of International Public Affairs and serves on numerous advisory committees. From 2020 to 2022, she was Director of the LSE Centre for Women, Peace and Security and in 2022 she became the first Andrew Mellon Fellow at UMass, Amherst. She has 25 years experience working globally on conflicts, crises, violent extremism and peacebuilding. At ICAN she spearheads the Women's Alliance for Security Leadership (WASL) comprising women-led peacebuilding organizations in 40 countries affected by war and violent extremism. She has led groundbreaking initiatives in scholarship, policy and practice. She was a drafter of UN Security Council Resolution 1325 and has led multi-country missions and studies on: women's contributions to peace processes including in South Africa, Colombia, Palestine; assessing the roles and vulnerabilities of men in fragile contexts such as Kenya, Jamaica, Liberia; and gender dimensions of violent extremism in Pakistan, Nigeria and elsewhere. She led a UN mission into Nepal's Maoist cantonments and served as the UN Standby Team of Mediation Experts first Senior Expert on Gender & Inclusion working on Somalia, Syria, Yemen and Libya. In 2020 in recognition for her services to international peacebuilding and women's rights, she was awarded an MBE in HM The Queen's New Year's Honors' List.

Major Kelly Atkinson is an Assistant Professor of Political Science at the US Air Force Academy, a Major in the US Air Force, and a career intelligence officer. She is a Council on Foreign Relations Term Member and a Chief of

Staff of the Air Force Strategic PhD Fellow. Kelly holds a PhD in Political Science and Women's Studies from the Ohio State University. Her doctoral research explored the effect of gender-based development policies on child recruitment into conflict, supported by archival research and interviews at the United Nations Headquarters in New York and the United Nations High Commissioner for Refugees in Geneva. Kelly has published multiple peer-reviewed journal articles on the role of refugee women in peacebuilding efforts within conflict dynamics. She is a Women in International Security 2021 Gender, Peace & Security Next Generation Symposium Fellow. At the Air Force Academy Kelly conducts research and teaches courses at the intersection of diversity and security.

Dr Whitney Grespin has worked in contingency contracting, professional military education, and international development on five continents. She received her PhD from King's College London at the UK Joint Services Command and Staff College, is a Lecturer at American University, and Director at Sloan Manor Consulting. Previously, she worked as an embedded advisor to the Somali Ministry of Defense, before which she supported the U.S. Army War College's Peacekeeping and Stability Operations Institute (PKSOI) as a Peace Operations Analyst and lead for the U.S. Army's Women, Peace and Security agenda. Whitney is a Term Member at the Council on Foreign Relations and is a member of the United Nation's Security Sector Reform Advisory Network.

Dr Alexis Henshaw is an Assistant Professor in Political Science and International Relations at Troy University. She is the author of *Why Women Rebel* (2017), *Digital Frontiers in Gender and Security* (2023), and is co-author of *Insurgent Women* (2019). Her work on gender and international security has been published in outlets including International Feminist Journal of Politics, Studies in Conflict and Terrorism, Journal of Global Security Studies, and Small Wars and Insurgencies.

Dr An Jacobs is a Senior Lecturer at Nottingham Trent University and The Royal Military Academy Sandhurst. She has operational experience as a Political Advisor for EULEX Kosovo and has conducted fieldwork in the Democratic Republic of the Congo, Burundi, Kenya, and Ukraine. Her research interests include peacebuilding, security sector reform, the EU's Common Security and Defence Policy, education as a peacebuilding tool, and the securitisation of migration and borders.

Changwook Ju is a PhD Candidate in Political Science at Yale University. He studies international relations and security. Before Yale, Changwook earned his MPP from the University of Chicago Harris School of Public Policy and dual undergraduate degrees in public policy and political science from Sungkyunkwan University in Seoul, South Korea. From 2011 to 2013, he served in the Republic of Korea Marine Corps, attaining the rank of sergeant.

Dr Katerina Krulisova is a Senior Lecturer at Nottingham Trent University. Katerina's research focuses on gender and security, specifically the UN's Women, Peace and Security Agenda, and women as perpetrators of political violence. Katerina serves as an Independent Advisor to the Czech Working Group on Women, Peace and Security.

Dr Robin Lovell is a senior scientist with more than 17 years of experience in environmental resource protection and conservation and has a proven track record of leadership and technical expertise in large, complex environmental projects including restoration, remediation, design, construction, and operations management. Robin has managed large teams of interdisciplinary experts to protect natural resources and ensure regulatory and legislative compliance. Currently, Dr Lovell is an Assistant Professor at Manhattan College focusing on the intersection of climate change and gender and has worked internationally in Southeast Asia and Europe solving problems in sustainable food production, the energy transition, and water resource management. Dr Lovell also teaches undergraduate courses in sustainability, geographic information systems (GIS), gender, established the Geography program at Manhattan College, and has published in internationally recognized, highly ranked journals.

Major Steve Maguire is in the British Army. Prior to joining the Army he completed a Masters in Intelligence and Security studies at the University of Salford, specialising in intelligence support to peacekeeping operations. He commissioned into the Royal Irish Regiment in 2011 and has served across Europe and Afghanistan. At the time of writing he is serving with the UK Land Warfare Centre with a responsibility for operational lessons in Europe. This includes an interest in the women, peace, and security agenda and assisting British Army thinking towards United Nations operations alongside the study of warfare. His interest in the women, peace, and security agenda is founded in academic study but also witnessing the need for gendered perspectives to be successful on military operations. Steve is also the Senior Land Editor at a professional military education website, *The Wavell Room*, and encourages junior military personnel to find their voice and write their ideas.

Dr Spencer B. Meredith, III is a Fulbright Scholar and Professor of National Security Strategy with over two decades experience researching, advising, and supporting democratic governance around the world. He is a subject matter expert in Russian, European, Central Asian and Middle Eastern politics, with dozens of conference presentations, peer-reviewed, and professional publications. His specializations include Special Operations and Irregular Warfare, and governance as a primary arena for geostrategic competition. Dr Meredith also serves as Strategic Advisor for the U.S. Joint Special Operations Command, a subject matter expert for the U.S. Army Special Operations Command, the lead governance curriculum designer for the U.S. Army Civil Affairs Qualification Course, and regular guest lecturer on Russian and Chinese strategic approaches to numerous Special Operations Subordinate Commands and Special Mission Units. He has contributed to multiple US government interagency working groups and strategic assessments over the past decade.

Colonel (Ret.) Jody M. Prescott was an active-duty judge advocate for 25 years. He had operational assignments in Bosnia as the Implementation Force claims chief for Sarajevo in the summer of 1996, and as the International Security Assistance Force chief legal advisor, 2008-2009. Col (Ret.) Prescott was an assistant professor at the U.S. Army Command and General Staff college, 2000-2003, and the U.S. Military Academy, 2009-2011. He has worked for the U.S. Department of Homeland Security as an attorney since retiring from the U.S. Army in 2011, and he is a lecturer at the University of Vermont teaching cybersecurity law & policy and energy & climate law. He is the author of *Armed Conflict, Women and Climate Change* (2018) and *Empirical Assessment in IHL Education and Training: Better Protection for Civilians and Detainees* (2021).

FOREWORD

Effective military operations require understanding the human terrain. As military practitioners, we are only too aware that civilians are disproportionately impacted by contemporary armed conflict, with women, children, and other vulnerable groups among the most affected. The implementation of the Women, Peace and Security (WPS) agenda is core to UK defence, and we are centralising it in our training of young officers at the Royal Military Academy Sandhurst (RMAS) through innovative academic and practical exercises. Understanding how contemporary conflict affects men, women, boys, and girls in different ways is essential for the sound and solid character of leadership we instil at RMAS.

In 2019, the UK Ministry of Defence adopted JSP1325 *Human Security in Military Operations* which emphasised the importance in developing a more profound understanding of the human terrain in which we conduct our operations. This Joint Services Publication sought to provide guidance on how military planners can better integrate WPS, children affected by armed conflict, human trafficking, and protection of civilians into their operations and planning. Four years on, it is worth reflecting on some of the challenges and opportunities, successes, and failures, and lessons learned. This ties in with the MoD's publication in December 2021 of JSP 985 *Human Security in Defence* which further embeds WPS within a broader Human Security framework. To exemplify this, a key annex is included in this collection. The document was prepared by SO1 Human Security/Civ-Mil Interaction Lt Col David Kemmis-Betty and succinctly illustrates the UK's approach, which is currently being conducted through the Human Security

Cell of 77 Brigade. This highlights and reiterates that WPS and military operations are not mutually exclusive. Rather, they reinforce each other. Integrating WPS in military operations is essential; it is right morally and ethically, and it also increases operational effectiveness.

Sandhurst is a unique forum in which we can bring together academics and practitioners, both policy and military, to discuss these challenges in a frank and open manner. Examining the integration of WPS into contemporary UK and international military operations requires asking ourselves tough questions, and we welcome these from the widest possible audience. Though there is much that the British Army and wider MoD can take pride in when it comes to integrating WPS, more still needs to be achieved, as some of the chapters included in this collection rightly highlight.

Looking to the future and how the British Army and the MoD can support the aspirations of the WPS agenda remains ever more crucial, if the UK is going to realise its global ambitions and meet the challenges of the contemporary operating environment. Critically examining what we do and how we are doing it is vital if we are to improve and become more operationally effective. This edited collection not only points out these challenges but offers substantive recommendations for military practitioners and policymakers. As such, it should be required reading for security professionals.

Major General ZR Stenning OBE
Commandant
Royal Military Academy Sandhurst

PREFACE

The adoption of United Nations Security Council resolution 1325 (UNSCR 1325) on 31 October 2000 led to the establishment of the Women, Peace and Security agenda (WPS). This resolution is based on four core pillars: prevention, participation, protection, relief and recovery, all underpinned by gender mainstreaming. Through these pillars, UNSCR 1325 "reaffirms the important role of women in conflict prevention and resolution, peace negotiations, peacebuilding, peacekeeping, humanitarian response and post-conflict reconstruction. It stresses the importance of women's equal participation and full involvement in all efforts to maintain and promote peace and security." At its core, the WPS agenda is concerned by the inordinate impact of violent conflict and war, especially conflict related sexual violence (CRSV), on women and girls and recognises the crucial, but often unacknowledged, role that women should, and do, play in conflict prevention, conflict management, and conflict transformation.

UNSCR 1325 has a significant impact on the way militaries can and should conduct their operations. However, militaries have been slow to adopt these ideas and adapt their behaviours to account for the importance of protecting civilians, gender mainstreaming, women's participation, responsibilities on children in conflict, and preventing egregious crimes, such as modern slavery and human trafficking. This book is a conversation between academics, civil society, and military practitioners to better understand the contemporary challenges and obligations militaries have towards realising the WPS agenda. Drawing on expertise from military, academia, and civil society it offers insights and recommendations on how militaries can effectively integrate the WPS agenda into their operations.

This collection follows an annual symposium organized by the department of Defence and International Affairs (DIA), held at the Royal

Military Academy Sandhurst (RMAS) on the theme: **Practicing/Employing/Working/Doing/Making Security: the Women, Peace and Security agenda and Human Security in Military Operations**. The event was attended by a wide array of military personnel, academics, and civil society groups and generated a conversation between academics and practitioners to better understand the contemporary challenges and obligations militaries have towards: protecting civilians; increasing women's role in peacekeeping and peacemaking; preventing the six grave violations against children; and stopping modern slavery and human trafficking. This conversation continues in this edited collection: *Women, Peace and Security in Military Operations*.

Addressing the challenges and opportunities of integrating human security and WPS into strategic and defence policy is made more fruitful by broad engagement between the variety of stakeholders that *do security* on the ground. The symposium sought to consider key questions including:

- The renewed relevance of WPS in 2022-23.
- How are women, children, and other vulnerable groups affected by conflict and what can be done to protect them?
- Can WPS and human security adapt in light of changes in the conduct of warfare, including the deliberate targeting of women and children and the use of human shields in urban operations?
- What are the best cases of successful integration of WPS and human security into strategic and defence policies and how can these be shared globally?
- How can theory best inform practitioners, and how can practitioners experiences be included in theory?
- What have been the major shortcomings of militaries globally and what can be done to address them?
- What role can and should militaries have in stopping human trafficking and supporting domestic police forces?
- How can human security and WPS address other vulnerable groups such as LGBTQI+, undocumented, and displaced communities?
- How can professional military education and trainers adapt to address the importance of gender mainstreaming and the human terrain in contemporary conflict?

Several of these questions are addressed explicitly or implicitly in this collection which seeks to explore how military operations can and should be more gender aware and how gender mainstreaming contributes to both gender equality and operational effectiveness.

The symposium was the first of the Sandhurst Trends in Conflict (STIC) series held at RMAS since the COVID-19 pandemic and included both an in-person event reflecting how global events shape not only policy, but also academic practice. It occurred not only against the backdrop of COVID-19, but major global shifts and geopolitical reconfigurations including: Russia's illegal invasion of Ukraine; the urgency of climate change; the growth of on-line extremism and radicalization, and the fallout from the withdrawal from Afghanistan. Each of these have significant impacts on women, children, and vulnerable groups; they highlight the pertinence of the WPS agenda and of the chapters included in this volume. Taken together these chapters offer a range of insights and recommendations for the renewal of the WPS agenda and its integration into the range of activities that constitute contemporary military operations.

This book starts with a thought-provoking piece by the keynote speaker, Sanam Naraghi-Anderlini MBE. Drawing on her decades of practical experience in working with women's groups and in drafting UNSCR 1325, she situates the agenda in 2023 and asks questions about what democratic countries that profess to be committed to the realization of gender equality and women's inclusion can and must do to revive the WPS agenda after the catastrophe of the Taliban's takeover of Afghanistan. Exploring the WPS agenda and its renewed relevance, she proffers some recommendations about what must be done to revitalize it. It frames the urgency of this moment and the chapters that follow it.

In chapter 1, An Jacobs and Katerina Krulisova offer an overview of the most salient contemporary debates and challenges for the effective conduct of gender mainstreaming in peace operations. Through highlighting major theoretical critiques of the agenda, they offer not only a clear understanding of these debates but highlight next steps for developing a better approach to bridging academia and practice. In chapter 2, Spencer Meredith provides a historic overview of women's participation in governance through military government and civil affairs operations. Highlighting the evolution of women's participation, it situates the 2017 US Women, Peace, and Security Act and highlights the opportunities and urgency of encouraging women's participation in partner states.

The book then looks at recent developments in Western militaries. In chapter 3, Steve Maguire provides an academic analysis of the current state of WPS in UN peacekeeping. He also draws on his personal experience in the British Army to comment on how WPS has contributed to both conceptual and doctrinal change, concluding with recommendations that will allow the British Army to lead in the implementation of the WPS agenda and increase its operational effectiveness. In chapter 4, Kelly Atkinson offers a substantive comparison of two US military operationalization(s) of human security: "hearts and minds" counterinsurgency and WPS. The first she argues is transactional whereas the second has potential to be transformational. She argues that for the WPS to succeed where COIN failed, a deeper understanding of the dynamics of human security is necessary. It concludes with a series of recommendations for the US DoD.

In chapter 5, Changwook Ju explores the impact of the presence of women combatants on wartime rape. It offers a statistical analysis that seeks to address the disproportionate number of "zero" observations in wartime rape data in existing studies. By accounting for zero observations, he finds that the presence of women combatants in state armed forces exacerbates the occurrence of wartime rape whereas it dampens it for rebel groups. The implications of these findings are of crucial importance for ensuring that protection against conflict related sexual violence is realised.

Following this, the book then turns to methods militaries can use to further understand the tactical and operational environments. In chapter 6, Whitney Grespin suggests the need to apply the PMESII-PT tool to improve partner nations' understanding of the operational environment and how best to include gender perspectives in their policies. Drawing on her experience working in Djibouti, she suggests that gender analysis should be used to contextualise the impacts of conflict on women and other marginalised groups. In chapter 7, Jody Prescott and Robin Lovell propose a Socio-Ecological Systems model to further understanding of the how armed conflict, gender inequality, and climate change exacerbate each other in different contexts. Their overlay model offers unique insights as to how militaries can build on other modelling tools such as PMESII-PT analysis to better prepare and plan operations in order to mitigate the gendered consequences of armed conflict and climate change.

Looking to the role of the Internet and the cyber domain on WPS, in chapter 8, Alexis Henshaw argues that the implementation of WPS should extend to the cyber domain since it is a source of gendered security threats and provides opportunities for increased participation of women. Looking at the four pillars of WPS, she suggests ways for policymakers and academics to engage and make the agenda relevant in the contemporary information environment.

In the concluding chapter, Max Thompson and Andree-Anne Melancon offer some insights on effective ways of integrating WPS into professional military education and training. Sharing their experiences as senior lecturers at RMAS, they offer recommendations on best practice for ensuring that military academies do not simply 'add WPS and stir', but rather integrate WPS into all areas of practical and academic training. They proffer a Sandhurst Approach as a way of realising the demands for effective training and education that may constitute best practice for other military academies around the world.

The importance of training and education is a constant theme in all the chapters of this book and ensuring that future practitioners understand that the importance, obligations, and the urgency of WPS is essential for its realisation in military operations. The undeniable human cost of conflict highlights the urgency with which the women, peace and security agenda ought to be integrated and operationalised not only in military operations, but across all governmental activity.

The editors would like to thank and acknowledge all speakers, chairs and attendees who took part in the symposium both those whose contributions make up this collection and those whose ideas helped inform their arguments. We would like to thank former Commandant Major General Duncan Capps for opening the symposium. We would also like to thank Brigadier (retired) Ian Thomas OBE, Dean of Academic Studies, for closing the symposium and his financial and practical support in enabling it to take place. Finally, we extend our thanks to Kirstin Howgate for her support for both the symposium and this collection.

ABBREVIATIONS

CAR	Conflict Analysis Report
COIN	counterinsurgency
CRSV	conflict related sexual violence
DIA	Defence and International Affairs (RMAS)
DoD	Department of Defense (US)
FAD	Forces Armées Djiboutiennes
FARC	Fuerzas Armades Revolucionarias de Colombia/ Revolutionary Armed Forces of Colombia
FCAS	fragile and conflict-affected states
FDCO	Foreign, Commonwealth and Development Office
GAPS	Gender Action for Peace and Security
GII	Gender Inequality Index
GWOT	Global War on Terror
HSIA	Human Security Individual Assessment
ICT	information and communications technology
JSP	Joint Services Publication
LGBTQI+	Lesbian, gay, bisexual, transgender, queer and intersex
LOAC	Laws of Armed Conflict
LOE	Lines of Effort
MMR	Maternal Mortality Ratio
MoD	Ministry of Defence
NAP	National Action Plan
OCdts	officer cadets
OMGUS	Office of Military Government, US
OP	ordered probit
P/CVE	preventing/countering violent extremism

PMESII-PT	Political, Military, Economic, Social, Information, Infrastructure, Physical Environment, and Time
PME	professional military education
PSVI	Preventing Sexual Violence Initiative
RAP	Regional Action Plans
RMAS	Royal Military Academy Sandhurst
SEA	sexual abuse and exploitation
SES	socio-ecological systems
STIC	Sandhurst Trends in International Conflict series
UN	United Nations
UNSCR	UN Security Council Resolution
WFP	World Food Programme
WPS	Women, Peace and Security
ZIOP	zero-inflated ordered probit

WOMEN, PEACE AND SECURITY

What Happened in Afghanistan Isn't Staying in Afghanistan

Sanam Naraghi Anderlini, MBE

In August 2021, the United States ended its 20-year occupation of Afghanistan with the most significant military withdrawal in modern times. From the outset, the White House promoted a rosy narrative. As President Biden claimed it was the "greatest airlift in history,"[1] a British former Royal Marines commando, with Prime Minister Johnson's knowledge and apparent approval, evacuated 200 Afghan cats and dogs. Meanwhile countless Afghan men and especially women – who had risked their lives to be police and army officers, judges, translators, reporters, peacebuilders, and human rights advocates, were left behind.

With the withdrawal, the Biden administration hoped to end the "forever war",[2] assuming that by putting the episode in America's rear-view mirror, 'what happened in Afghanistan, would stay in Afghanistan'. Washington's think-tanks also framed the debate in simplistic binary terms of being for or against continued military presence, and assumed the issues were limited to Afghanistan. But assumptions are not fact, and what happened in Afghanistan always had wider geopolitical significance. For years prior to the withdrawal, Afghan women and those of us working internationally had warned against the Taliban's intent and its implications.

As women peacebuilders and rights advocates, we believed military withdrawal was necessary, but it had to be conducted as part of a clear overarching political vision and a diplomatic strategy. A central message was the importance of ensuring Afghan women – especially those with

1 The White House, "Remarks by President Biden on the End of the War in Afghanistan," 31 August 2021, https://www.whitehousegov/briefing-room/speeches-remarks/2021/08/31/remarks-by-president-biden-on-the-end-of-the-war-in-afghanistan/

2 The White House, "Remarks by President Biden on the End of the War in Afghanistan".

expertise in peacebuilding, rights, and security matters – were invited as an independent delegation to negotiations about their own, and Afghanistan's future.

This call was a key pillar of UN Security Council Resolution (UNSCR) 1325 in 2000 and the subsequent women, peace and security (WPS) agenda, that emerged as a result of a global grassroots-based campaign in the late 1990s. This introductory chapter discusses the significance of the WPS agenda in the context of Afghanistan focusing on the missed opportunities for realizing the agenda, the geopolitical implications of the Taliban's takeover for peace and security internationally, and implications for the future of WPS. It concludes with some recommendations on immediate actions that the UK and other governments should take to reassert their core commitments and credibility internationally, vis-à-vis peace, security and upholding universal human rights values.

Afghanistan and WPS

The WPS agenda and the plight of Afghan women were intertwined from the very start. The Taliban's coercion and erasure of Afghan women from public life in the 1990s during their first period of control, was among the drivers of the WPS agenda. In 2001, in the aftermath of 9/11, the United States claimed the war on the Taliban was to free women from their oppression.[3] In Bonn 2001, there was a consistent demand by, and for the inclusion of Afghan women in the negotiations that led to the establishment of the Afghan state. For the years that ensued, Afghan women persisted in claiming their space in the political, peace and security decision making arenas of their country.

Despite many setbacks they had significant successes. Women's rights were protected under the constitution. A parliamentary quota of minimum 27 per cent of seats in the lower house and 17 per cent in the upper house, ensured the presence of women in political spaces and government positions. Girls' schools and higher education became possible. In 2003 just 6 per cent of girls attained higher education schooling. By 2018 the ratio

3 Laura Bush, "Laura Bush on Taliban Oppression of Women," *The Washington Post*, 17 November 2001, https://www.washingtonpost.com/wp-srv/nation/specials/attacked/transcripts/laurabushtext_111701.html.

had risen to 40 per cent.[4] Countless women-led civil society organizations opened across the country. Women were visible in every arena of public life including the media, sports, and the economy. A special judicial court was established to address cases of violence against women, with hundreds of lawyers and judges fanning across the country to prosecute and bring perpetrators to justice.

The risks were always profound. With the NATO military presence, violence was a constant threat. Women in public life, especially those who challenged the Taliban's regressive ideology or spoke against corruption in politics, were always targeted. But they took on the risks, because they understood how much was at stake, for themselves and future generations. It was, thus not surprising that in 2013 when the United States sought Qatar's assistance in facilitating dialogue with the Taliban, Afghan women raised the alarm bells. When the Qatari government invited the Taliban to establish its offices in Doha, and the dialogues began, women's rights advocates and peacebuilders mobilized again to demand a seat at the negotiations.

In 2017, President Trump signed the bipartisan *Women, Peace, and Security Act* into law. The Act calls on the US government to ensure the inclusion of women in peace processes, where the US has a role. But between 2013 and 2021, when Qatar hosted negotiations between the United States and the Taliban and opened doors to diplomatic delegations from the United Nations, Europe, and other countries across the region, Afghan women peacebuilders were never invited as an independent delegation. Despite US legislation calling for the inclusion of women in the peace process, the US envoy, Zalmay Khalilzad repeatedly dismissed the demand and stonewalled Afghan women. This deliberate exclusion of women's independent delegations to the actual talks contravenes the word and spirit of the ten UN security council resolutions that make up the WPS agenda, the countless national action plans that governments, like the UK, have adopted, and the US WPS Act itself. The exclusion bolsters the argument that states are not living up to their own commitments.

After significant pressure, concessions were made. The Afghan government included four women in its political delegation. Norway hosted events for Afghan women. But throughout the years, the participation of

4 Grant Farr, "Female Education in Afghanistan after the Return of the Taliban," *E-International Relations*, 23 April 2022, https://www.e-ir.info/2022/04/23/female-education-in-afghanistan-after-the-return-of-the-taliban/

women was relegated to UN Women and its partnerships with international NGOs, such as Mina's List. They put much time and effort into convening separate women's conferences producing important recommendations for the official talks. But they were disconnected from the most significant negotiations. Afghan women were literal and metaphorical side-shows, even though their lives and futures were being negotiated away.

Meanwhile as the talks continued in Doha, the Taliban's targeting of Afghan civilians including the bombing of a maternity hospital,[5] girls' schools and university,[6] continued. These acts of terror were clear indications of the movement's ideology and intent. However, the international community ignored them. To compound the issues, the Trump administration agreed to the release of 5000 Taliban prisoners in 2020. Yet again the voices, warnings, and concerns of Afghan women were ignored. In 2021 the Taliban triumphed. Taking control of the entire country, parading in cities wearing the remnants of military equipment left behind by the NATO forces. By September 2022, Afghan girls had been excluded from school for over a year. Women were kept out of work and forced into begging. Incidences of forced marriage are rising. The plight of the 3797 women police officers is largely unknown. In short, the analysis and warnings of Afghan women were proven true as Kabul fell.

Betraying Afghan Women and the Implications for States that Champion the WPS Agenda

For the wider WPS agenda, the implications of what happened in Afghanistan and to Afghan women are profound. It is especially so, given that many of the western countries that have championed the WPS agenda were key actors in Afghanistan. Norway and the United Kingdom are key examples. The UK has always proudly proclaimed to be the 'pen holder' on WPS at the UN Security Council. Since 2008 it has championed the prevention of sexual violence in conflict pillar of the agenda. In 2020 it was the first member state to publicly pledge support for the 'protection of women peacebuilders'. As former State Minister James Cleverly (Foreign

5 BBC, "Afghan Attack: Maternity Ward Death Toll Climbs to 24," *BBC*, 13 May 2020, https://www.bbc.co.uk/news/world-asia-52642503.

6 France 24, "Death Toll Tops 50 in Suicide Attack on Female Afghan Students, UN Says," *France 24*, 3 October 2022, https://www.france24.com/en/asia-pacific/20221003-death-toll-rises-in-afghan-suicide-attack-on-female-students-in-exam-centre.

Secretary at the time of writing) stated to the UN Security Council on October 29, 2020:

> *The voices of women peacebuilders must be heard and amplified. But for many, speaking up and carrying out their vital work comes at great personal risk. No woman should have to risk her safety to heal her community.*
>
> *That is why I am proud that the UK has supported the International Civil Society Action Network (ICAN) in developing the Protection Framework for Women Peacebuilders.*
>
> *This vital piece of work provides guidance to states and multilateral organizations on how best to prevent and respond to reprisals.*
>
> *I am pleased to announce that the UK will formally commit to supporting this framework and its recommendations. I urge others to do the same.*

In August 2021, however, the commitment was not kept, and the lofty promises to "protect", "hear", and "amplify the voices" of women peacebuilders were broken. Afghan women police and security officers, lawyers, judges, those who had served in government, as well as those who had worked in their communities to build peace, foster justice and mitigate violence were left behind. Apart from international civil society and independent military personnel and government officials who took on the responsibility to assist, protect and evacuate, there was no systemic formal effort to enable their security or evacuation.

This failure to uphold and enact twenty years of WPS policies and national action plan pledges, reveals three hard truths. First, despite the rhetoric, the policies and commitments have been skin-deep, and disconnected within the machinery of government. In effect, one unit of the Foreign, Commonwealth and Development Office may make a commitment to protect women peacebuilders, but other sections and other ministries such as the Home Office and immigration services may renege on that commitment, with impunity.

Second the failure to stand with Afghan women is a blow to the credibility of, and trust in, the UK and multilateral institutions. The UK cannot claim to uphold and value human rights and women's rights, if political leadership and diplomatic corps adhere to the whims and demands of adversaries whose ideology is inherently misogynistic. In other words, the Taliban might seek to erase women from public life, but our institutions and governments must not emulate them or sink to their standards.

The third is more complex. In its withdrawal from Afghanistan, the US may have hoped that it would be the end of the so-called "forever war".

But a year on, a new reality is emerging. The Taliban, as the ideologically extremist David's-defeat of the hyper-powered but unfocused Goliath of NATO, triggered three, inter-related tectonic geopolitical shifts: the resurgence of transnational violent extremism, the rise of state-based authoritarianism, and the further demise of multilateralism and human rights norms. This is not the end of the forever war, rather the prologue and harbinger of a new and darker era globally. It is imperative to be aware of the signs and triggers of this change.

1. The first trigger, the Taliban's victory gave a new lease on life to transnational violent extremist movements weaponizing faith, race, and ethno-nationalism around the world. So-called Jihadi movements weaponizing Islamist identity are applauding and inspired by the Taliban. They are showing new muscle, clearly evident in the attacks on women and girls elsewhere as in the Philippines where a law passed to prohibit child marriages is being challenged by the extreme Islamic religious leaders, who reference the Taliban's rule. Claiming child marriage adheres with Islamic laws, they are calling for the lifting of the ban. This is not an isolated case, and deeply alarming. The Biden administration triggered a global green light on gender-based violence. While Afghan women and girls are most impacted, women worldwide are paying the price for the resurgent extremism.
2. The second trigger has been a rise of state-based authoritarianism. From Putin's war on Ukraine to Erdogan's demands in NATO, leaders with dictatorial tendencies have been emboldened. The illusion of power and unity once projected by the US and its western allies is shattered. The Ukraine war may have repaired some damage, but the international credibility of the US is tattered. The rise of authoritarianism has further undermined democracy and rights globally. Granted, that 9/11 and the subsequent 'War on Terror' era that prompted the occupation of Afghanistan and Iraq, had already accelerated the decline of human rights. Under the guise of countering terrorism, many states have shut down civil society, human rights, and amped up their security apparatus. But the US decision to circumvent the elected Afghan government to negotiate and comply with the demands of the clearly unelected Taliban, even as they bombed maternity clinics and schools, sent the world a clear signal: The US doesn't stand by its stated principles.

3. The shredding of what remained of the rule-based order, multilateralism and the already fragile human rights norms and systems, is the third and final blow. This too dates to the 'War on Terror' and the impunity with which the US and NATO forces killed civilians in Afghanistan, Iraq, Libya, Syria, Yemen and laid siege to international laws and norms. But the handling of Afghanistan – from the unrepresentative nature of the Doha negotiations to the treatment of Afghans – especially Afghan women – during and after the August withdrawal, lowered the bar. Few world leaders can credibly claim to champion women's rights or a feminist foreign policy when the help to Afghan women police officers, lawyers and rights activists has been so limited. In effect, by ignoring and abandoning Afghan women, the US and its allies not only lost in Afghanistan but accelerated the decline of the already-shaky post-war global order that was inclusive of a broad swathe of nations. These developments have a magnified impact in contexts where people are already overwhelmed with the challenges of ongoing conflict and extremism.

The US and NATO's withdrawal from Afghanistan was meant to end the "forever war" and enable reductions in military expenditure. Instead in 2022 global military expenditure topped $2 trillion and a new arms race has started. Looking forward, the likelihood of a new cold war is rising with the US and its traditional western allies (including Australia and New Zealand) on one side, and China, Russia and their authoritarian client states to other. Attention to Africa, Asia, the Middle East and Latin America is either waning or will be increasingly more transactional. However, in this globalized world, the issues of human and state security are more interrelated than ever before, making our multilateral systems and spaces for diplomacy and cooperation, even more necessary.

In reality, the post 1945 institutions that were established to foster peace and security internationally, have not been reformed adequately to be fit for purpose in the 21st. The implications of climate change and environmental degradation are also becoming more evident and more extreme, with the threat of flooding, droughts and other human-induced natural emergencies rising and fueling additional conflict.[7] While the COVID-19 pandemic may

7 United Nations Framework Convention on Climate Change, "Conflict and Climate," July 12, 2022, https://unfccc.int/blog/conflict-and-climate.

be ebbing, the lack of international solidarity and support to communities in fragile contexts and competition over vaccines further shredded trust and relations between wealthier and poorer nations. In 2022, 24 per cent of the world's population lives in fragile contexts.[8] In November 2021 the UN Office for the Coordination of Humanitarian Affairs (OCHA) estimated that 274 million people would need humanitarian assistance in 2022 (a 17 per cent increase on 2021). Their target was to reach 183 million people, with a fund-raising goal of $41 billion.[9] By mid-year, only $22 billion had been pledged. By the third quarter of 2022, the number of people in need of aid had risen to 324 million, while the funds received was only $19.2 billion.[10]

The net result is sobering. Peace can no longer be taken for granted anywhere, and international systems of aid are woefully inadequate. Economic inequalities within and between states are also heightened and exacerbating fundamental trust and social cohesion. Globally, peacefulness within countries has dropped by an average of 2 per cent per year over the past 15 years, according to the 2021 *Global Peace Index*. Today only 3 per cent of the world's population live in countries with "open" conditions for civil society action.[11] While resources are desperately needed to address humanitarian, development and peacebuilding needs, the opposite is occurring. Governments are reducing overseas development aid budgets but increasing their defense and military expenditure.

From the outset in 2000, the WPS agenda that called for greater inclusivity in crisis and conflict prevention and greater attention to the gendered dimensions of conflict, was offering new solutions to address the challenges of the 21st. But in 22 years, despite much rhetorical commitment and policymaking, little systemic and operational change is evident. A key consequence is that the warnings, solutions and strategies of women and youth at the frontlines of such crises is not heeded or supported. A profound consequence is that people in the poorest regions including women and young people in local communities are often the first and most severely impacted.

---

8 Organisation for Economic Co-operation and Development, "States of Fragility 2020," September 19, 2022, https://www.oecd.org/dac/states-of-fragility-fa5a6770-en.htm.

9 United Nations Office for the Coordination of Humanitarian Affairs, "At a Glance – Global Humanitarian Overview," accessed November 28, 2022, https://gho.unocha.org/intro/glance.

10 United Nations Office for the Coordination of Humanitarian Affairs, "Global Humanitarian Overview 2022 – Humanitarian InSight," accessed November 28, 2022, https://hum-insight.info/?_gl=1.

11 CIVICUS, "Quick Facts – CIVICUS – Tracking Conditions for Citizen Action," June 22, 2022, https://monitor.civicus.org/quickfacts/.

Yet, even when faced with threats and pressures, women peacebuilders continue to take on the responsibility to protect their families and communities. There is also significant collective action. Countless global and localized vibrant social movements ranging from climate action to racial justice, women's rights and LGBTQI+ rights are harnessing the power of technology to connect for action. In the past twenty years, civil society has also engaged more effectively with state institutions. The WPS agenda is a prime example of the potential transformation that can arise when states and the UN system heed and support the vision and demands of civil society. There are now a number of civil society-based networks that are tackling the myriad challenges our societies face. If reinforced, sustained and enabled, they have the potential to foster a bottom-up ecosystem for peace, resilience, equality, and pluralism that is locally rooted and globally connected.

Women-led civil society organizations and women peacebuilders are critical actors at each of these levels, with their participation significantly improving the sustainability of peace and reconciliation processes. Yet despite the solutions they offer, women continue to be absent from relevant national or international decision-making spaces. Their absence risks the loss of valuable cultural and traditional conflict resolution approaches.

At the May 2022 ICAN forum *Redefining Security, Reclaiming Power*, women peacebuilders that have been ICAN's partners in the Women's Alliance for Security Leadership (WASL) representing over 40 countries worldwide, identified these challenges. But they also had clear response:[12]

1. While cognizant of the scale of complexity and threat facing their societies, they felt compelled to maintain their activism to foster peace, address needs and provide positive alternatives for their own communities; and
2. The solidarity, connectivity, and collaboration they have within WASL and in relation to ICAN are essential to enabling their work, creativity, and wellbeing.

In other words, the WPS agenda is even more relevant for the future, but it requires governments to significantly recalibrate current business as usual.

12 International Civil Society Action Network (ICAN), *2022 Forum Findings*, forthcoming December 2022, available at www.icanpeacework.org

What Next for Afghanistan and What Next for WPS?

Most urgently, the international community must do right by the Afghan women and men who are at high risk. Those inside country who are resisting the Taliban, need to be supported. Some are (re)forming civil society organizations to fight to reassert their basic rights, ranging from access to schools and universities, to healthcare, livelihoods and public presence, and protection against violence domestically or practices such as forced marriage, that leads to a life of sexual servitude.

The following recommendations are relevant not only to Afghanistan but other contexts of conflict and fragility.

1. Locally rooted women-led civil society organizations must have safety to operate and be supported financially as well as for organizational and technical development. They are the engine for the protection and progress of rights for women.
2. Local women peacebuilders must be systematically heard and heeded in international fora.
3. Countries convening dialogues with the Taliban – on humanitarian aid, security, economics, or politics – must include women in their own diplomatic teams and ensure that independent Afghan women's delegations (from local organizations) also participate.
4. Humanitarian aid to Afghanistan must be gender responsive and sensitive to issues of marginalization and violence that communities such as the Hazara experience. Afghan women's organizations and women in communities must be included in the design and delivery of aid to ensure it reaches the most marginalized.
5. Governments must also unite in pressing the Taliban to restore rights to education, livelihoods, healthcare, protection against violence, and freedom of movement. Grand statements by Taliban leaders in social media calling for 'girls' education' are not sufficient. There is an urgent need for cooperation with neighboring and regional states to identify the leverage points that can influence the Taliban.
6. Direct cash transfer to families (through remittances or charity donations) must be supported and enabled. Afghans should have safe and unfettered access to their bank accounts, savings and services such as Western Union.

7. The security and protection of women to work in government in the legal, judicial, and security sectors, and those active in civil society (media, arts, education and health), and receive and retain salaries must be guaranteed in line with the Taliban's own stated commitments (pt.8).[13]
8. The Taliban's conditions/demands that women must have a Mahram to go to work, access public life and services, is untenable, and a de facto prison sentence for countless women/girls – especially female headed households. The UN and donors must not accept such restrictions.
9. Governments and lawmakers must expedite support for the evacuation of Afghan women most at risk, and still in hiding in the country, and the resettlement of those languishing in transit humanitarian camps. In the context of the UK, a coordination working group comprising representatives from the Foreign, Commonwealth and Development Office, Home Office, Parliament and relevant civil society organizations (those engaged with at-risk Afghans), could expedite their identification, verification, and transit.

We must also have the courage to reflect on the harsh lessons and failures of the Afghan experience. We need to heed the calls of women in other contexts, like Yemen and Libya, and avoid the mistakes from Afghanistan. By negotiating separately with the Taliban, the US legitimized an extremist movement that has limited support within Afghanistan. By creating a two-party 'Intra-Afghan' process that is more focused on political (and economic) power sharing than on actual peace, the US signaled that violence is the ticket to the political arena. By not holding the Taliban or other parties (that may be part of the government's delegation) to account for the ongoing violence, the US and the international community enabled their impunity. The Taliban intensified its campaign of targeted killings of especially women and the levels of violence rose by 25 per cent between 2019 and 2020 while the negotiations were taking place. With an office in Doha, passports and visas, the Taliban was given the same legitimacy as a democratic nonviolent group, despite the atrocities they carried out.

---

13 Ahmad Seir *et al.*, "Taliban vow to respect women, despite history of oppression," AP News, 18 August 2021, https://apnews.com/article/afghanistan-taliban-kabul-1d4b052ccef113adc8dc94f965ff23c7.

Concurrently, the international community ignored the fact that 50 per cent of Afghans are children (under 18), and 51 per cent of the population are female. Violence perpetrated in communities is thus knowingly targeting women (femicide), children and educational institutions.

It is time for a strategic and geopolitical reboot. In the Cold War, the US instigated its share of proxy wars and fronted many a dictator, but the hard power came hand in hand with fostering the ideals of democracy, pluralism, and equal rights. These aspirations are alive now. From Sudan, Myanmar, and Palestine to Afghanistan where over 62 per cent of the population is under 25, young people and women risk their lives daily, to bring these values to life. They need solidarity, recognition, protection, and resources to do their vital work. Russia, China, and other authoritarian states like Turkey, or Arab Gulf states with their veneer of modernity are antithetical to them. The credibility of the US, the UK, and other democracies may be severely tarnished because of the wars in Afghanistan, Iraq, Libya, and Yemen. However, in the face of rising extremism and authoritarianism, democratic countries need a course correction and to remember, compared to other powers, where our strengths and unique vantage points lie.

To prevent the forever war turning into war forever, instead of an arms race and trillion-dollar defense spending, now more than ever we need to stand for and practice the principles of universal human rights, justice, and pluralism. This is not naivete. Nor is helping Afghans an act of benevolence. It is, as one former soldier said to me, an act of integrity to salvage the soul of America and other still-democratic nations.

References

BBC, "Afghan Attack: Maternity Ward Death Toll Climbs to 24," *BBC*, 13 May 2020, https://www.bbc.co.uk/news/world-asia-52642503.

Bush, Laura. "Laura Bush on Taliban Oppression of Women," *The Washington Post*, 17 November 2001, https://www.washingtonpost.com/wp-srv/nation/specials/attacked/transcripts/laurabushtext_111701.html.

CIVICUS, "Quick Facts – CIVICUS – Tracking Conditions for Citizen Action," 22 June 2022, https://monitor.civicus.org/quickfacts/.

Cleverly, James. "Ensuring Women's Full Participation in Peace Processes: Statement to the UN Security Council" (Open Debate on Women, Peace and Security, 29 October 2020), https://www.gov.uk/government/speeches/ensuring-womens-full-participation-in-peace-processes.

Farr, Grant. "Female Education in Afghanistan after the Return of the Taliban," *E-International Relations*, April 23, 2022, https://www.e-ir.info/2022/04/23/female-education-in-afghanistan-after-the-return-of-the-taliban/

France 24, "Death Toll Tops 50 in Suicide Attack on Female Afghan Students, UN Says," *France 24*, October 3, 2022, https://www.france24.com/en/asia-pacific/20221003-death-toll-rises-in-afghan-suicide-attack-on-female-students-in-exam-centre.

International Civil Society Action Network (ICAN), *2022 Forum Findings*, forthcoming December 2022, available at www.icanpeacework.org

Organization for Economic Co-operation and Development, "States of Fragility 2020," 19 September 2022, https://www.oecd.org/dac/states-of-fragility-fa5a6770-en.htm.

Seir, Ahmad *et al.*, "Taliban vow to respect women, despite history of oppression," AP News, 18 August 2021, https://apnews.com/article/afghanistan-taliban-kabul-1d4b052ccef113adc8dc94f965ff23c7.

The White House, "Remarks by President Biden on the End of the War in Afghanistan," 31 August 2021, https://www.whitehouse.gov/briefing-room/speeches-remarks/2021/08/31/remarks-by-president-biden-on-the-end-of-the-war-in-afghanistan/

United Nations Framework Convention on Climate Change, "Conflict and Climate," 12 July 2022, https://unfccc.int/blog/conflict-and-climate.

United Nations Office for the Coordination of Humanitarian Affairs, "At a Glance – Global Humanitarian Overview," https://gho.unocha.org/intro/glance.

United Nations Office for the Coordination of Humanitarian Affairs, "Global Humanitarian Overview 2022 – Humanitarian InSight," https://hum-insight.info/?_gl=1.

1

GENDER MAINSTREAMING IN PEACE OPERATIONS

An Jacobs and Katerina Krulisova

The gendered nature of conflicts and the need for transformative action to achieve more peaceful and just societies are globally accepted. There exists a plethora of resolutions, directives, action plans, and operational mandates that highlight the importance of gender equality in both the practice and theory of peace operations. As the international community reflects on 23 years of the Women, Peace and Security (WPS) agenda, academics, activists, and practitioners recognise its unfulfilled promises. One of the often-criticised aspects of security governance is gender mainstreaming. The term 'gender mainstreaming' gained international prominence in the 1980s. While early efforts at mainstreaming focused primarily on increasing the number of women in public life, gender mainstreaming is much more than just adding women into existing power structures. Gender mainstreaming denotes the practical steps to be taken to achieve more gender-equal, gender-sensitive, and just environments.

The academic community points to several issues with both the theory and practice of gender mainstreaming.[1] Whether it concerns the fuzziness of the term or the unclear impact of mainstreaming initiatives, feminist security scholarship offers productive insights that can help security practitioners to better understand gender mainstreaming and to implement it more effectively. This chapter focuses on some of the most pressing issues

---

1 Karin Landgren, "WPS and Gender Mainstreaming", In *The Oxford Handbook of Women, Peace, and Security*. 2018; Rahel Kunz, "Messy feminist knowledge politics: a double reading of post-conflict gender mainstreaming in Liberia", *International Feminist Journal of Politics* 22, no. 1 (2020): 63-85; Maria-Adriana Deiana and Kenneth McDonagh, "'It is important, but...': translating the Women Peace and Security (WPS) Agenda into the planning of EU peacekeeping missions", *Peacebuilding* 6, no. 1 (2018): 34-48; Vanessa F. Newby and Clotilde Sebag, "Gender sidestreaming? Analysing gender mainstreaming in national militaries and international peacekeeping", *European Journal of International Security* 6, no. 2 (2021): 148-170.

at the forefront of the debate in the study of gender mainstreaming in peace operations.[2] The successes and challenges of implementation of gender mainstreaming in these complex operational environments are studied by scholars across academic disciplines. While it is beyond the scope of this chapter to comprehensively review these discussions, it will cover some of the most salient debates and insights that are valuable to both academics and practitioners, with a view to encouraging enhanced communication between the two.

Our review consists of four main parts. First, we discuss definitions, and we address institutional and policy changes in the context of gender mainstreaming. We then engage with the conceptual discussions, before exploring the key issues with the implementation and operationalisation of gender mainstreaming in peace operations. The final part of the chapter discusses the role pre-deployment training and continuous professional development can play to tackle some of the challenges raised. We conclude by summarising the state of the discipline and providing pointers for further research.

Gender Mainstreaming: Definitions and Institutional Developments

Gender mainstreaming appears to be an often-repeated buzzword that (not only) policymakers use. Daly notes that gender mainstreaming is widely understood as "the most 'modern' approach to gender equality" and one of the "essentially technical concepts that has managed to achieve a relatively wide currency in a short time."[3] Krook and True go further to argue that gender mainstreaming has become an international norm that is now widespread and firmly institutionalised.[4]

Gender mainstreaming is often seen as a practical tool for the achievement of gender equality. In its simplest meaning, gender

2 We understand peace operations broadly, including efforts to prevent violent conflict, diffuse a violent situation, and build long-term sustainable peace (please see "What is a Peace Operation", Center for International Peace Operations (ZIF), https://www.zif-berlin.org/en/what-peace-operation).

3 Mary Daly, "Gender mainstreaming in theory and practice", *Social Politics: International Studies in Gender, State & Society*, 12, no. 3 (2005): 433-450, 433.

4 Mona Lena Krook and Jacqui True, "Rethinking the life cycles of international norms: The United Nations and the global promotion of gender equality." *European Journal of International Relations*, 18, no. 1 (2012): 103-127, 105.

mainstreaming considers gendered implications of any public policy. Probably the most often cited definition of gender mainstreaming is in the UN Economic and Social Council 1997 Agreed Conclusions. It states that gender mainstreaming is "the process of assessing the implications for women and men of any planned action, including legislation, policies or programmes, in all areas and at all levels".[5,6] Daly adds that the distinctiveness of gender mainstreaming lies within its aim to "*institutionalize* equality by embedding gender-sensitive practices and norms into the structures, processes, and environment" of any public policy-making body.[7]

The concept of gender mainstreaming was introduced during the Nairobi World Conference on Women in 1985. The gender mainstreaming revolution began at the United Nations Fourth World Conference on Women in Beijing in 1995, which called for the integration of a gender perspective into all activities and areas.[8] Gender mainstreaming became an integral part of the UN's WPS agenda. The first WPS Resolution, UN Security Council Resolution (UNSCR) 1325 (2000) recognized the gendered nature of conflicts and calls for greater participation of women in peace processes, conflict resolution and peacebuilding.[9] It also recalled the Windhoek Declaration and the Namibia Plan of Action on Mainstreaming a Gender Perspective

5 The whole definition reads: 'Mainstreaming a gender perspective is the process of assessing the implications for women and men of any planned action, including legislation, policies or programmes, in all areas and at all levels. It is a strategy for making women's as well as men's concerns and experiences an integral dimension of the design, implementation, monitoring and evaluation of policies and programmes in all political, economic and societal spheres so that women and men benefit equally and inequality is not perpetuated. The ultimate goal is to achieve gender equality.' United Nations, ECOSOC Agreed Conclusions 1997/2.

6 International organisations and national institutions often follow the key points set out in the ECOSOC definition. For example, UK JSP 985 defines gender mainstreaming as: "implications for women and men of any planned action, in all areas and at all levels, in order to assure that the concerns and experiences of both sexes are taken into account." Ministry of Defense, Joint Service Publication 985, Human Security in Defence, 2021: vi (for more information see Annex 1). In their WPS Action Plan, NATO states the following: "NATO as a political and military alliance contributes to the implementation [of WPS] by systematically integrating gender perspectives into planning and execution of operations, training, exercises, and policies, as well as dialogue and partnerships." NATO/EAPC Women, Peace and Security Policy and Action Plan, 2018, 10.

7 Daly, "Gender mainstreaming in theory and practice", 435, emphasis added by authors.

8 Jacqui True, "How effective is gender mainstreaming in international peace and security policymaking?", in *Handbook on Gender in World Politics,* ed. Jill Steans and Daniela Tepe-Belfrage (Cheltenham: Edward Elgar Publishing, 2016).

9 For more discussion on feminist work on Resolution 1325 see: Sheri Lynn Gibbings, "No angry women at the United Nations: political dreams and the cultural politics of United Nations Security Council Resolution 1325," *International Feminist Journal of Politics,* 13, no. 4 (2011): 522-538; Carol Harrington, "Resolution 1325 and post-Cold War feminist politics", *International Feminist Journal of Politics,* 13, no. 4 (2011): 557-575.

in Multidimensional Peace Support Operations.[10] Further resolutions have expanded on the need for women's active participation in peacekeeping and peacebuilding and focus on sexual violence in armed conflicts.

The commitment to gender mainstreaming is repeated and re-affirmed throughout the WPS-related policy documentation. All member states of the UN are tasked with the development of a National Action Plan (NAP) that detail the specific measures and actions taken to implement UNSCR 1325. At the time of writing, 103 countries have adopted a NAP.[11] On top of this, there are also several Regional Action Plans (RAPs). International organisations such as the African Union, ECOWAS, League of Arab States, the European Union, and NATO all have RAPs in place.[12] Gender mainstreaming has also been written into mission mandates[13], UN Peacebuilding Commission documentation[14], and policies of the UN's Department of Peacekeeping Operations.[15]

Despite all this effort, scholars have identified ongoing challenges to gender mainstreaming in peace operations. Several researchers point to the disappointing impact of WPS and gender mainstreaming initiatives. Basu, Kirby and Shepherd have described WPS as "a vast normative infrastructure of significant ambition but questionable impact."[16] Terms

---

10 Nina J. Lahoud, "What Fueled the Far-Reaching Impact of the Windhoek Declaration and Namibia Plan of Action as a Milestone for Gender Mainstreaming in UN Peace Support Operations and Where Is Implementation 20 Years Later?", *Journal of International Peacekeeping*, 24, no. 1-2 (2020): 1-52.

11 Updated list is available online at Peacewomen.org. "1325 National Action Plans (NAPs)." Available online at: http://1325naps.peacewomen.org/.

12 For more information see: WPS Focal Points Network. "Regional Action Plans." Available online at: https://wpsfocalpointsnetwork.org/regional-action-plans/.

13 Anne-Kathrin Kreft, "The gender mainstreaming gap: Security Council resolution 1325 and UN peacekeeping mandates", *International Peacekeeping*, 24, no. 1 (2017): 132-158.

14 Laura J. Shepherd, "Victims of violence or agents of change? Representations of women in UN peacebuilding discourse", *Peacebuilding*, 4, no. 2 (2016): 121-135; Laura J. Shepherd, *Gender, UN Peacebuilding, and the Politics of Space: Locating Legitimacy*. (Oxford: Oxford University Press, 2017).

15 United Nations, Department of Peacekeeping Operations, Department of Field Support, "Gender Responsive United Nations Peacekeeping Operations." Available online at: https://peacekeeping.un.org/sites/default/files/english_gender_responsive_united_nations_peacekeeping_operations_policy_1.pdf.

16 Soumita Basu, Paul Kirby, and Laura J. Shepherd, "Women, Peace and Security: a critical cartography", in *New Directions in Women, Peace and Security*, ed Soumita Basu, Paul Kirby, and Laura J. Shepherd, (Bristol, Bristol University Press, 2020): 1-25, 3.

such as "unfulfilled mandate"[17] or "gender sidestreaming"[18] have been used to capture the ongoing challenges of implementation of gender mainstreaming in peace operations. The following section introduces the key academic criticism on the theory and practice of gender mainstreaming in peace operations. We start by focusing on three major conceptual debates: the equation of gender with women; the 'add women and stir' practice; and the need for decolonization of gender mainstreaming.

Conceptual Debates

There is a vast amount of academic debate on gender mainstreaming in peace operations, but three key theoretical critiques stand out. These are: 1) The conflation of sex and gender 2) The 'add women and stir' thinking; and 3) The postcolonial feminist critique of WPS and peacebuilding. At the heart of all these issues lies either a misunderstanding of or resistance to key feminist principles. This is a seemingly paradoxical situation; after all, gender mainstreaming and WPS are the results of decades-long and continuing feminist activism.[19] At the heart of this activism is the idea of uprooting patriarchal and racist power structures and institutions. However, instead, gender mainstreaming and WPS are often co-opted into business-as-usual, integrated into existing patriarchal power structures.[20] What follows below is the academic critique of some of the key building blocks of this nexus.

---

17 Jacqui True, "The Unfulfilled Mandate-Gender Mainstreaming and UN Peace Operations", *Georgetown Journal of International Affairs*, 10, Issue 2 (2009): 41-50.

18 Newby and Sebag define sidestreaming as "the practice, deliberate or unintentional, of sidelining women and relegating them to specialised spaces in international peace and security while attempting gender mainstreaming or increased gender integration.' Vanessa F. Newby and Clotilde Sebag, Gender sidestreaming? Analysing gender mainstreaming in national militaries and international peacekeeping", *European Journal of International Security*, 6, no. 2 (2021): 148-170, 149.

19 Sanam Naraghi Anderlini, "Civil society's leadership in adopting 1325 resolution", in *The Oxford Handbook of Women, Peace, and Security*, ed. Sara E. Davies, and Jacqui True (Oxford: Oxford University Press, 2019), 38.

20 Audrey Reeves, "Feminist knowledge and emerging governmentality in UN peacekeeping: Patterns of co-optation and empowerment", *International Feminist Journal of Politics*, 14, no. 3 (2012): 348-369.

Gender ≠ Women

Feminist literature often points out the conceptual conflation of the terms sex and gender. Sex denotes a biological category, whereas gender is commonly understood as an identity category and/or power relation.[21] The identity category or social construction understanding denotes the meanings particular people ascribe to femininities and masculinities. Gender is much more than "a set of ideas about male and female people", but "a way of categorizing, ordering, and symbolizing power, of hierarchically structuring relationships among different categories of people, and different human activities symbolically associated with masculinity or femininity."[22] Understanding gender as a power relation or power structure acknowledges that this social construction of meaning dictates unequal power relationships between people, which has wide implications for anyone affected by the peace operation.

One of the most often repeated critiques of current peacebuilding discourses and practices is the conflation of the term gender with sex and subsequently, gender with women.[23] Instead of power relations, peacekeepers/peacebuilders often understand the term gender as a synonym for women. Charlesworth notes that this is deeply embedded in the institutional heritage of gender mainstreaming at the UN, which was at the time focused on women's lives and international development.[24] Similarly, the language of *Women*, Peace and Security lands meaning to the initiatives as women focused. These issues are, of course, extremely important. Women's experiences of conflicts matter, their agency and participation in peace processes are key, and recognition that their voices are not always heard or listened to is progress. Yet, women's rights mainstreaming is *not* the same as gender mainstreaming.

Gender mainstreaming requires gender analysis. The ongoing analysis of complex power relationships is often missing from contemporary initiatives.[25] Without the intersectional analysis of power inequalities,

21 Shepherd, *Gender, UN Peacebuilding, and the Politics of Space.*

22 Carol Cohn. *Women and wars: Contested histories, uncertain futures,* (Hoboken: John Wiley & Sons, 2013), 3.

23 Heidi Hudson, "Decolonizing the mainstreaming of gender in peacebuilding: toward an agenda for Africa." *African Peacebuilding Network (APN) Working Paper* 8 (2016); Shepherd, *Gender, UN Peacebuilding, and the Politics of Space.*

24 Hilary Charlesworth, "Not waving but Drowning: Gender Mainstreaming and Human Rights in the United Nations", 18 HARV." Hum. Rts. J 1 (2005).

25 Daly, "Gender mainstreaming in theory and practice".

gender mainstreaming loses much of its potential. Tracing the power dynamics requires an analyst to resist the temptation to treat women as a single category of agency-less victims. Critical questions need to be asked about the differences in experiences of women, men, girls, and boys. Any good intersectional analysis should also focus on the experience of LGBTQI+ people in local contexts[26] as well as race, class,[27] urban-rural status, and religion. Without taking into consideration the variety ways in which conflict impacts people with different identities and experiences, positive peace cannot be achieved.[28]

Add Women and Stir

Closely related to the conflation of gender with sex, and subsequently gender with women is the 'add women and stir' or 'add women and hope' critique.[29] This refers to the strategy of simply adding women into existing institutions without considering the structural inequalities that underpin and govern them. Such a strategy of 'gender-balancing' is tokenistic and insufficient.[30] Peace operations must move beyond this logic and work actively towards transforming the institutional cultures and practices. On top of this, several studies show that while there is a positive trend in the participation of women in peace operations since 2000, overall progress in gender-balancing has been slow.[31]

---

26 Jamie J. Hagen, "Queering women, peace and security." *International Affairs,* 92, no. 2 (2016): 313-332; Lisa Davis and Jessica Stern. "WPS and LGBTI rights", in *The Oxford Handbook of Women, Peace, and Security,* eds Sara E. Davies and Jacqui True (Oxford: Oxford University Press, 2018).

27 María Martín de Almagro, "Producing participants: Gender, race, class, and women, peace and security." *Global Society,* 32, no. 4 (2018): 395-414.

28 For detailed discussions on positive and negative peace, please see Johan Galtung, & Dietrich Fischer, "Positive and Negative Peace", in *Springer Briefs on Pioneers in Science and Practice,* ed Johan Galtung., vol 5. Springer, Berlin.

29 See for example: Jasmine-Kim Westendorf, "'Add women and stir': the Regional Assistance Mission to Solomon Islands and Australia's implementation of United Nations Security Council Resolution 1325", *Australian Journal of International Affairs,* 67, no. 4 (2013): 456-474; Sabrina Karim and Marsha Henry. "Gender and peacekeeping." In *The Oxford Handbook of Gender and Conflict,* eds. Fionnuala Ní Aoláin, Naomi Cahn, Dina Francesca Haynes and Nahla Valji. (Oxford: Oxford University Press, 2018); Kenneth McDonagh and Maria-Adriana Deiana. "Add women and hope? Assessing the gender impact of EU common security and defence policy missions: a policy report." (Dublin: Dublin City University, 2017).

30 UN Department of Peacekeeping Operations (DPKO) defines gender balancing as "the degree to which men and women hold the full range of positions in a society or organisation". Natasja Rupesinghe, Eli Stamnes, and John Karlsrud, "WPS and female peacekeepers", in *The Oxford Handbook of Women, Peace, and Security,* eds Sara E. Davies and Jacqui True (Oxford: Oxford University Press, 2018), 206.

31 Rupesinghe, Stamnes, and Karlsrud, "WPS and female peacekeepers".

Equal representation is key and should be an explicit goal. Karim and Beardsley note that the concept of gender-balancing was meant to be a rights-based approach, stating that "women [...] should be able to participate in the same activities as men and have adequate representation".[32] Yet, most policy-makers instead focus merely on the functionality of a peace operation.[33] Adding women, ideally a "few kind women",[34] is seen as necessary for increasing the effectiveness of peace operations in general[35] or as a deterrent to sexual abuse and exploitation (SEA) by both local men and male peacekeepers.[36] This logic does not take into consideration broader gendered culture in peace operations,[37] militarised/peacekeeping masculinities,[38] or the intersectional nature of peacekeeping economies.[39]

The effectiveness arguments centre around the role of the woman peacekeeper/peacebuilder as more effective in an often stereotypical, binary and essentialist understanding of gender roles; typical feminine characteristics desired in a peacekeeper/peacebuilder are patience, caring,

32 Sabrina Karim and Kyle Beardsley, "Female peacekeepers and gender balancing: Token gestures or informed policymaking?." *International Interactions*, 39, no. 4 (2013): 461-488, 465.
33 Robert Egnell, Petter Hojem, and Hannes Berts, *Gender, military effectiveness, and organizational change: the Swedish model* (Berlin: Springer, 2014).
34 Johanna Valenius, "A few kind women: gender essentialism and Nordic peacekeeping operations", *International peacekeeping*, 14, no. 4 (2007): 510-523.
35 Donna Bridges and Debbie Horsfall, "Increasing operational effectiveness in UN peacekeeping: Toward a gender-balanced force." *Armed Forces & Society* 36, no. 1 (2009): 120-130.
36 For more details on SEA in peace operations see: Olivera Simić, "Does the presence of women really matter? Towards combating male sexual violence in peacekeeping operations", *International Peacekeeping*, 17, no. 2 (2010): 188-199; Jasmine-Kim Westendorf and Louise Searle, "Sexual exploitation and abuse in peace operations: trends, policy responses and future directions", *International Affairs*, 93, no. 2 (2017): 365-387; Sarah Smith, "Accountability and sexual exploitation and abuse in peace operations", *Australian Journal of International Affairs*, 71, no. 4 (2017): 405-422; Stephen Moncrief, "Military socialization, disciplinary culture, and sexual violence in UN peacekeeping operations", *Journal of Peace Research*, 54, no. 5 (2017): 715-730.
37 Helena Carreiras, "Gendered culture in peacekeeping operations." *International Peacekeeping*, 17, no. 4 (2010): 471-485.
38 See: Claire Duncanson, "Forces for good? Narratives of military masculinity in peacekeeping operations", *International Feminist Journal of Politics*, 11, no. 1 (2009): 63-80; Claire Duncanson, Forces for good?: Military masculinities and peacebuilding in Afghanistan and Iraq. (Berlin: Springer, 2013); Annica Kronsell, "Sexed bodies and military masculinities: Gender path dependence in EU's common security and defense policy." *Men and Masculinities*, 19, no. 3 (2016): 311-336; Sandra Whitworth. Men, militarism, and UN peacekeeping: A gendered analysis (Boulder: Lynne Rienner Publishers, 2004); David Duriesmith, *Masculinity and New War: The gendered dynamics of contemporary armed conflict* (London: Routledge, 2016).
39 Kathleen M Jennings and Morten Bøås, "Transactions and interactions: Everyday life in the peacekeeping economy." *Journal of Intervention and Statebuilding*, 9, no. 3 (2015): 281-295.

peacefulness and empathy.[40] Women are expected to be "the 'peaceful and friendly face' in peace operations".[41] Their roles are to better connect with local women, engage and support women and children, and create a greater sense of safety and security for local populations.[42] In areas with a high prevalence of sexual or gender-based violence, the lack of women police officers creates a barrier to achieving justice as "local women are more likely to report crimes to women police".[43] On top of this, the presence of women is expected to have a civilizing effect on the men peacekeeper/ peacebuilder and deter the perpetration of sexual abuse and exploitation against local populations.[44] Wilén argues that such stereotypes create an "added burden" that female personnel carries and in effect undermine gender equality.[45]

Existing research, however, paints rather a mixed picture of this perceived unique contribution. Karamé's findings indicate that local populations focus on the 'uniform' of the peacekeeper rather than their gender.[46] Heinecken adds that women peacekeepers are in no way equipped to make a difference 'as women' since they do not receive "specialist training in terms of how to interact with the local population, gather intelligence or deal with cases of sexual violence."[47] In relation to changing mission cultures from within, women tend to be absorbed into the often hyper-masculine and militarised mission settings rather than being able to effectively challenge them.[48] Jennings argues that women peacekeepers are often unwilling to report SEA committed by male staff

---

40 Lesley J Pruitt, "All-female police contingents: Feminism and the discourse of armed protection", *International Peacekeeping*, 20, no. 1 (2013): 67-79, 67; see also: Lindy Heinecken, "Are Women 'Really' Making a Unique Contribution to Peacekeeping?: The Rhetoric and the Reality", *Journal of International Peacekeeping*, 19, no. 3-4 (2015): 227-248.

41 Nicole Jenne and Fiorella Ulloa Bisshopp, "Female Peacekeepers: UNSC Resolution 1325 and the Persistence of Gender Stereotypes in the Chilean Armed Forces", *International Peacekeeping*, 28, no. 1 (2021): 134-159, 143.

42 Pruitt, "All-female police contingents", 67.

43 Marie Vlachová and Lea Biason, eds. *Women in an insecure world: violence against women: facts, figures and analysis*. DCAF; 2005; cited in Pruitt, "All-female police contingents", 68.

44 Gerard J. DeGroot, "A few good women: Gender stereotypes, the military and peacekeeping", *International Peacekeeping*, 8, no. 2 (2001): 23-38; See also: Bernd Beber, Michael J. Gilligan, Jenny Guardado, and Sabrina Karim. "Peacekeeping, compliance with international norms, and transactional sex in Monrovia, Liberia", *International Organization*, 71, no. 1 (2017): 1-30.

45 Nina Wilén, "Female peacekeepers' added burden", *International Affairs*, 96, no. 6 (2020): 1585-1602.

46 Kari H. Karamé, "Military women in peace operations: Experiences of the Norwegian battalion in UNIFIL 1978–98", *International Peacekeeping*, 8, no. 2 (2001): 85-96.

47 Heinecken, "Are Women 'Really' Making a Unique Contribution to Peacekeeping?", 247.

48 Simić, "Does the presence of women really matter?".

against locals.[49] Karim and Beardsley find correlation between the inclusion of female peacekeepers *and* personnel from countries with good records on gender equality correlates with "lower level SEA allegations reported against military contingents."[50] This evidence strongly suggests that instead of adopting the essentialist logic add women, stir and hope for the best, peace operations require introspective gender analysis and comprehensive cultural and institutional rethinking.

Decolonising Gender Mainstreaming

Debates on both peace operations and WPS point to the persistent need to critically question and challenge colonial legacies and logic that continue to shape contemporary practices. Hall highlights that the term 'postcolonial' does not denote a simple before and after chronology; instead, problems caused by colonialism are not resolved and violence, which is often gendered, has not ceased.[51] Legacies of colonialism continue to significantly shape hierarchies of gender, race, class, sex, disability, etc. Attention must be paid to the ways in which colonial power relations are reproduced in both discourses and practices of peace operations.

There is a myriad of ways in which coloniality continues to shape global governance.[52] Fernández and Guerra argue that "the Eurocentric, colonial, and racist gazes often inform peacebuilding imaginaries."[53] A similar critique is levelled against the current state of WPS. Post-colonial feminist scholars argue that the WPS agenda is rooted in colonial mindsets and practices. Martin de Almagro argues that WPS policy is "part of a liberal peacebuilding framework that is racialised, patriarchal, classist, heteronormative and Western-centric at its core."[54] Parashar adds that WPS

49 Kathleen M. Jennings, "Unintended consequences of intimacy: political economies of peacekeeping and sex tourism", *International Peacekeeping*, 17, no. 2 (2010): 229-243.
50 Sabrina Karim and Kyle Beardsley. "Explaining sexual exploitation and abuse in peacekeeping missions: The role of female peacekeepers and gender equality in contributing countries", *Journal of Peace Research*, 53, no. 1 (2016): 100-115, 100.
51 Stuart Hall, *Essential essays, volume 1: Foundations of cultural studies* (Durham: Duke University Press, 2018).
52 Meera Sabaratnam, "Avatars of Eurocentrism in the Critique of the Liberal Peace", *Security Dialogue*, 44, no. 3 (2013): 259-278; Oliver P. Richmond and Roger Mac Ginty, "Where now for the critique of the liberal peace?", *Cooperation and Conflict*, 50, no. 2 (2015): 171-189.
53 Marta Fernández and Lucas Guerra. "Peacebuilding and Postcolonial Subject." *The Palgrave Encyclopedia of Peace and Conflict Studies, eds.* Oliver Richmond and Gëzim Visoka (Springer, 2020), 1-12, 1.
54 María Martín de Almagro, "Producing participants: Gender, race, class, and women, peace and security", *Global Society*, 32, no. 4 (2018): 395-414, 142.

endorses "the neoliberal projects of Western governments".[55] Haastrup and Hagen argue that WPS "is steeped in racialized hierarchies manifested in whiteness", where whiteness denotes "a system that reifies the status quo of White privileges, protections, and material advantages over People of Color".[56]

This thinking reproduces the binary of the civilized 'West' [read rich, progressive, educated, liberal, gender equal] and backwards 'rest' [poor, uneducated, authoritarian, patriarchal].[57] In this view, the Global South becomes a mere passive recipient of gender norms; these norms are brought and implemented into local contexts by Western institutions.[58] 'West' becomes an efficient problem-solver of issues occurring 'elsewhere'.[59] Gender knowledge is then produced in the 'West' and gender expertise is claimed by individuals working for big international institutions rather than locals. Kunz *et al* show how some gender experts categorise gender issues on a scale of 'basic' or 'first-generation' such as female genital mutilation or HIV/AIDS that apply to local settings, and 'nuanced' or 'complex' such as childcare or labour market policies.[60] Deina and McDonagh note that EU peacebuilders tend to "associate questions of gender inequality with spaces other than Europe".[61] Such lack of self-reflection as well as an incomplete understanding of local context hinders effective workings of peace operations.

This thinking is also projected to the ways in which peacebuilders/ peacekeepers perform their own gender roles. Studying gender mainstreaming in EU missions, Kronsell discusses the construction of EU protector masculinity against a vulnerable femininity to be protected

55 Swati Parashar, "The WPS agenda." in *The Oxford Handbook of Women, Peace, and Security*, eds Sara E. Davies and Jacqui True (Oxford: Oxford University Press, 2018).

56 Toni Haastrup and Jamie J. Hagen, "Global racial hierarchies and the limits of localization via National Action Plans", in *New Directions in Women, Peace and Security*, ed Soumita Basu, Paul Kirby, and Laura J. Shepherd, (Bristol, Bristol University Press, 2020): 133-151, 133-134.

57 For an excellent discussion on women's status and measures of 'civilisation' see: Ann Towns, "The status of women as a standard of 'civilization'", *European Journal of International Relations*, 15, no. 4 (2009): 681-706.

58 On translation of global norms in local contexts see: Zeynep Capan, Gulsah, Filipe Dos Reis, and Maj Grasten. *The Politics of Translation in International Relations* (Cham: Palgrave Macmillan, 2021).

59 Shepherd, "Victims of violence or agents of change?", 332.

60 Rahel Kunz, Elisabeth Prügl, and Hayley Thompson, "Gender expertise in global governance: Contesting the boundaries of a field", *European Journal of Politics and Gender*, 2, no. 1 (2019): 23-40, 33.

61 Maria-Adriana Deiana and Kenneth McDonagh, "Translating the women, peace and security agenda into EU common security and defence policy: Reflections from EU peacebuilding", *Global Society* 32, no. 4 (2018): 415-435, 430.

from aggressive and violent local men.[62] Yet, not all peacekeeping/ peacebuilding masculinities appear to be equal. In their study of the EU Rule of Law Mission, EULEX Kosovo, Deina and McDonagh note that peacebuilders consider 'Western Europeans' more progressive as opposed to more 'unprofessional' Eastern Europeans.[63] These complex gender, race and geopolitical hierarchies need to be further reflected upon in order to make gender mainstreaming more effective.

Hudson links the flaws of contemporary peacebuilding to the issues with the "liberal-feminist inspired gender-peacebuilding nexus" and proposes strategies to achieve more decolonial peacebuilding.[64] Decolonial peacebuilding should focus on "effecting structural and institutional change; conflict prevention and protection; justice and healing; development and capacity-building; and governmental and multilateral humanitarian efforts that respect the agency of those most affected".[65] Peace operations are highly political endeavours, not technical exercises devoid of existing power relations and hierarchies. Decoloniality requires that all involved in designing and implementing any peace operation reflexivity.[66] They should critically reflect not only on the local and international legacies of colonial structures, but also on their own positionality and relative privilege.[67]

The Implementation of Gender Mainstreaming in Peace Operations

For gender mainstreaming to have a meaningful impact on peacebuilding, successful implementation is vital. Several scholars have highlighted a

62 Kronsell, "Sexed bodies and military masculinities".

63 Deiana and McDonagh. "Translating the women, peace and security agenda into EU common security and defence policy", 431-432.

64 Hudson, "Decolonising gender and peacebuilding: feminist frontiers and border thinking in Africa", 195, 196.

65 Hudson, "Decolonising gender and peacebuilding: feminist frontiers and border thinking in Africa", 196.

66 On reflexivity see: Brooke Ackerly and Jacqui True, "Reflexivity in practice: Power and ethics in feminist research on international relations", *International Studies Review, 10*, no 4 (2018): 693-707; Cynthia Enloe, "Being Reflexively Feminist Shouldn't Be Easy." *Researching War. Feminist Methods, Ethics and Politics*, ed. Annick Wibben (London: Routledge 2016): 258-259.

67 For a discussion on peacekeeping economies and their effects, see Jennings, Kathleen M., and Morten Bøås. "Transactions and interactions: Everyday life in the peacekeeping economy." *Journal of Intervention and Statebuilding* 9, no. 3 (2015): 281-295; Henry, Marsha. "Parades, parties and pests: Contradictions of everyday life in peacekeeping economies." *Journal of Intervention and Statebuilding* 9, no. 3 (2015): 372-390.

'policy-implementation gap'; while there is ample discussion on gender issues in policy circles, the mainstreaming of gender in peacebuilding operations on the ground is lagging.[68] Institutional progress has been made, but the operational impact of gender mainstreaming remains questionable.

Early work on the implementation of gender mainstreaming focused on women (as opposed to gender) and engaged with questions on numbers and participation. This represented the 'zeitgeist' at the time, reflecting a largely numerical interpretation to gender mainstreaming, in line with the aforementioned 'add women and stir' approach. Olsson, for example, explored the number of women working for multidimensional peacekeeping operations, and how task divisions are gendered, with a significantly higher number of women represented in civilian components of peacekeeping compared to military aspects of peacekeeping.[69,70] In one of the first statistical studies on the topic, Karim and Beardsley examined women's representation and participation in military components of UN peacekeeping and found evidence that female military personnel are often deployed to lower-risk areas, and less to countries where there might be a greater need due to sexual violence and gender inequality.[71]

The approach of 'add women and stir' equates gender mainstreaming successes to having a certain percentage of women participating in operational activities in war-torn societies rather than understanding power relations that lead to gender inequality. In this regard, Nduka-Agwu explores complexities of gender mainstreaming in Liberia and Sierra Leone and explains that military peacekeepers are trained in a "hypermasculine milieux" which encourages displays of behaviour where gender is used to make sharp "distinctions between the sexes and where an aggressive and

---

68 Natalie Florea Hudson, "Engendering UN Peacekeeping Operations," *International Journal*, 60, no. 3 (2005): 785-807; Vanessa F. Newby and Clotilde Sebag, "Gender sidestreaming? Analysing gender mainstreaming in national militaries and international peacekeeping," *European Journal of International Security* 6, no. 2 (2021): 148-170; Louise Olsson and Theodora-Ismene Gizelis, "Advancing gender and peacekeeping research," *International Peacekeeping* 21, no. 4 (2014): 520-528; Jacqui True, "How Effective is Gender Mainstreaming in International Peace and Security Policymaking?." *In Handbook on Gender in World Politics*, Eds Jill Steans and Daniela Tepe, (Edward Elgar Publishing, 2016).

69 Louise Olsson. "Mainstreaming Gender in Multidimensional Peacekeeping: A Field Perspective," *International Peacekeeping* 7, no. 3 (2000): 1-16.

70 For UN targets on women in peace operations, please see https://peacekeeping.un.org/en/gender

71 Sabrina Karim and Kyle Beardsley. "Female peacekeepers and gender balancing: Token gestures or informed policymaking?," *International Interactions* 39, no. 4 (2013): 461.

often misogynist heterosexuality is celebrated".[72] The 'add women and stir' approach in gender mainstreaming therefore gives an incomplete and superficial depiction of the concept of gender instead of considering deeper layers of gender power relations that can lead to the perpetuation of a particular type of masculinity in the military context. Newby and Sebag have argued that gender mainstreaming needs fundamental military reform that disassociates "violence and combat skills from masculinity".[73] Instead of considering gender as a mere descriptor of an identity category, peacebuilders must 'do' gender in a way that identifies gender as a power relation, that helps us understand that peacebuilding is gendered.

Since the early 2000s, thinking about the implementation of gender mainstreaming in peacebuilding has moved beyond the numerical approach, now covering a wider range of aspects, themes, and challenges to implementation. A first reoccurring theme involves peacebuilders' attitudes towards gender mainstreaming, and how gender is understood by those deployed in conflict-affected environments. Research has shown that peacebuilders are often reluctant to accept gender as a priority, with a tendency to "see gender issues as secondary".[74] This reluctance to take gender seriously inevitably has an impact on the operationalization; gender mainstreaming is considered a box that needs to be ticked to satisfy the gender adviser, the chain of command or external actors, rather than something that requires implementation at all phases, levels, and activities of the mission. To overcome some of the behavioural challenges to gender, a more structured and continuous interface between practitioners in peacebuilding operations and feminist scholars could iron out some fundamental misunderstandings and transform mainstream approaches to security in relation to the WPS agenda.[75]

Considering gender mainstreaming as a box-ticking exercise reflects more deep-rooted challenges within the gender debates. First, as previously discussed, it echoes – and to some extent stems from – many peacebuilders' incorrect understanding of gender as women and the

---

72 Adibeli Nduka-Agwu "'Doing gender' after the war: Dealing with gender mainstreaming and sexual exploitation and abuse in UN peace support operations in Liberia and Sierra Leone." *Civil Wars* 11, no. 2 (2009): 184.

73 Newby and Sebag, "Gender Sidestreaming?", 148.

74 Deiana and McDonagh, "Translating the women, peace and security agenda into EU common security and defence policy", 423.

75 Laura Huber and Natalie F. Hudson, "Deepening the Conversation: Feminism, International Policing and the WPS Agenda," *International Peacekeeping* 26, no. 5 (2019): 579-604, 579.

gender agenda as only benefiting women and girls.[76] Second, it relates to the negative connotations many peacebuilders still have regarding the word 'gender'; it is often considered unwelcome or uncomfortable which 'should be avoided if possible'.[77] Third, it reflects the conceptual challenge that gender mainstreaming is often presented as a predetermined process, or something that can have a clear end-result.[78] This not only implies that once the objectives are achieved, the process is completed, but also – as discussed under the section on colonial thinking – that some cultures have already achieved this so-called end-result while others have not. In that sense, implementing gender mainstreaming has suffered from an 'us versus them' dichotomy, where gender equality is considered as something that has been achieved in Europe and the West and only requires work in other parts of the world. In EU missions, where the deployed staff is predominantly European, this results in the belief that internal, in-mission gender mainstreaming is not necessary.[79]

A second reoccurring theme in debates on the operationalization of gender mainstreaming in peacebuilding missions relates to how gender is mainstreamed within the structures and working practices of the mission itself. While progress has been made on the appointment of gender advisers and gender focal points,[80] successful implementation of gender mainstreaming still heavily relies on the work of *femocrats* present in the mission.[81] This indicates that to achieve positive institutional change and ensure gender is mainstreamed throughout the mission, champions of the gender agenda are needed in strategic operational positions. "Feminist actors working within institutions play an important role as they contribute to the production and diffusion of norms".[82] This is no different in the

76 True, "How effective is gender mainstreaming" 461.

77 An Jacobs and Katerina Krulisova, "Gender Mainstreaming in Peacebuilding," *LSE Centre for WPS*, December 2021, 9.

78 Shepherd, *Gender, UN Peacebuilding, and the Politics of Space,* 72-75.

79 Deiana and McDonagh, "Translating the women, peace and security agenda into EU common security and defence policy", 423.

80 Gender advisers provide guidance to mission management on introducing the gender perspective into all aspects of the mission. They are supported by gender focal points, located in different parts of the mission. "HQ SACT Office of the Gender Adviser", NATO, https://www.act.nato.int/gender-advisor.

81 Femocrats are 'feminists in government, whether in women's policy agencies or elsewhere, and whether at national, subnational, or transnational levels of governance.' Sawer, Marian. "Femocrat." *The Wiley Blackwell Encyclopedia of Gender and Sexuality Studies* (2016): 1-2.

82 Roberta Guerrina and Katharine AM Wright, "Gendering normative power Europe: lessons of the Women, Peace and Security agenda," *International affairs* 92, no. 2 (2016): 293-312, 297.

institutional microcosm of a peacebuilding mission.[83] In addition, visible support for the gender agenda from the mission senior management is a key factor in determining the attitudes towards gender issues, and hence affects successful implementation. Peacekeeping leadership has also been identified as one of four UN procedural strategies required for gender mainstreaming.[84] Gender advisers have indicated that having support from their senior mission management makes a substantial difference in mainstreaming gender across all levels, phases, and tasks of the mission.[85]

Discussions on concepts, institutions, and structures can overlook the important question of how gender mainstreaming can help achieve wider security and development goals.[86] The third and final trend identified in gender mainstreaming discussions therefore relates to its impact on the implementation of the mission mandate, and how gender mainstreaming can make a difference to the local context on the ground. After all, gender equality is a goal for both the internal structure of peace operations, as well as for the activities of the operation in the host nation.[87] Understanding the gender-related power dynamics on the ground through inclusive gender analysis contributes to a more complete understanding of the local context, enhanced situational awareness, and it enables operations to respond more effectively to societal needs. In fact, the potential impact of peacebuilding on local gender power-relations requires careful consideration to avoid undermining the route to equal peace, and to support the overall peace process.[88] Context-based understandings of gender could therefore enhance communication between the local and the international.[89]

Although it is widely acknowledged that gender mainstreaming is crucial to building positive peace, less is known about *how* gender mainstreaming can help achieve long-lasting peace in different societies,

83 Camille Conaway, Jolynn Shoemaker, Allison Adams-Alwine, and Jennifer Park, *Women in United Nations Peace Operations: Increasing the Leadership Opportunities* (Washington, DC: Georgetown University Press, 2008).
84 Hudson, "Engendering UN Peacekeeping Operations".
85 Jacobs and Krulisova, "Gender Mainstreaming".
86 True, "How effective is gender mainstreaming" 462.
87 Sahana Dharmapuri, "Why is Gender Equality in Peacekeeping Important" in *Not Just a Numbers Game. Increasing Women's Participation in UN Peacekeeping* (International Peace Institute, 2013).
88 Luise Olsson, *Gender equality and United Nations peace operations in Timor Leste* (Leiden: Martinus Nijhoff Publishers, 2009).
89 Heidi Hudson, "'Doing' security as though humans matter: A feminist perspective on gender and the politics of human security." *Security Dialogue* 36, no. 2 (2005): 155-174; Heidi Hudson, "It Matters How You 'Do' Gender in Peacebuilding: African Approaches and Challenges." *Insight on Africa* 13, no. 2 (2021): 142-159.

and what the key opportunities, weaknesses and threats of the process are. Operational activities in the host nation are often highly gendered and fail to consider different social needs.[90] Ansorg and Haastrup, for example, discuss the impact of the different societal contexts in Ukraine and Afghanistan on police reform. They find that the appointed Minister of Internal Affairs, Eka Zguladze, enabled femocrats to positively bring substantial and gender inclusive reform in the Ukrainian police, while societal constraints in Afghanistan impinge the implementation of positive gender norms.[91]

Engaging with in-depth studies of the impact of gender mainstreaming in specific operational realities helps to better understand the complexity faced by peacebuilding programmes in conflict-affected environments. It can provide insights not only regarding the impact of specific gender-related programmes in the local context, but also enhance our understanding of multifaceted local-international relationships and interactions.[92] Some scholars have pointed to the fact that peacebuilding measures are to a greater or lesser extent able to contribute to the transformation of masculinities in post-conflict settings.[93]

The Impact of Training on Gender Mainstreaming in Peace Operations.

Overcoming longstanding challenges related to the implementation of gender mainstreaming requires fundamental changes in culture and attitudes. Change can be long-term if achieved through processes of social learning.[94] Providing training that helps operational staff unlearn existing associations with terms like gender and feminism is key. This enables them to conduct appropriate gender analysis and enhance not only their understanding of what gender mainstreaming means but also incorporate

90 Helen SA Basini, "Gender mainstreaming unraveled: The case of DDRR in Liberia," *International Interactions* 39, no. 4 (2013): 535-557.
91 Nadine Ansorg, and Toni Haastrup. "Gender and the EU's Support for Security Sector Reform in Fragile Contexts," *JCMS: journal of common market studies* 56, no. 5 (2018): 1127-1143.
92 Pol Bargués-Pedreny and Maria Martin de Almagro, "Prevention from Afar: gendering resilience and sustaining hope in post-UNMIL Liberia," *Journal of Intervention and Statebuilding* 14, no. 3 (2020): 327-348.
93 Maike Messerschmidt and Hendrik Quest, "Peacebuilding measures and the transformation of masculinities: Looking at Liberia and Uganda," In *Gender Roles in Peace and Security* (Springer, Cham, 2020), 79-100.
94 Shirley M. Hord, and James L. Roussin, *Implementing change through learning: Concerns-based concepts, tools, and strategies for guiding change* (Corwin Press, 2013).

it in their daily work.[95] Effective gender training can serve as a tool that can transform individuals and institutions.

Colombia's peace education strategy serves as a good example of this.[96] By incorporating gender education into the school curriculum as part of the peace building process, teachers have acquired more knowledge on gender issues and are therefore better equipped to address gender inequalities in the classroom. This in turn has resulted in a more equal environment not only in spaces of learning, but also in the wider Colombian society. It gave students tools to think critically about gender practices in their communities, encouraging them to "realise the power game related to gender so that they can take action", and avoid these behaviours in their own societal environment. [97] However, despite the potential that training offers to engender positive change, two caveats need to be considered. First, training in itself is not sufficient to generate change, but instead should be one of many tools as part of a broader strategy. Secondly, appropriate strategies for gender equality are bound to be context-specific, which means there is no one-size-fix-all framework that can be applied across different societal contexts.

The potential long-term impact of gender training on the effectiveness of peace operations has been widely noted among academics and practitioners. There is particular attention for how 'gender' is problematised in training and might consequently influence attitudes and behaviours. Carson, for example, emphasises that gender training for Australian peacekeepers "does not problematise power relations between the sexes and is only covered as a way of understanding the peacekeeping context, and not in relation to the attitudes and behaviours of peacekeepers themselves".[98] Similarly, Holvikivi emphasises that the discourse used during gender training – varying from a narrow focus on women as victims to more expansive gender interpretations – contributes to the creation

95 Fiona Mackay and Georgina Waylen, "Introduction: Gendering "new" institutions," *Politics & Gender* 10, no. 4 (2014): 489-494.

96 For more details on the Colombian Peace Process, see Jorge L Fara-Zamora, Andrés Molina-Ochoa, Nancy C. Doubleday (eds), *The Colombian Peace Agreement. A Multidisciplinary Assessment* (Routledge, New York, 2021).

97 Andrea Luna Ruiz, "Gender Approach: A Requirement for Peacebuiding," *Global Journals of Human-Social Science Linguistics and Education* 19, no. 1 (2019): 23-30; UNICEF, "Learning for peace," 2016, Available at: https://eccnetwork.net/wp-content/uploads/Gender-Education-and-Peacebuilding-A-Review-of-Learning for-Peace-Case-Studies.pdf, 29.

98 Lisa Carson, "Pre-deployment 'gender' training and the lack thereof for Australian peacekeepers," *Australian Journal of International Affairs* 70, no. 3 (2016): 275-292, 283.

of a specific understanding of gender.[99] It is therefore vital to present gender as inclusive as possible, emphasising power relations and moral values. Holmes, analysing the training experiences of female Rwandan peacekeepers, sees an opportunity in the training space, as a "field of norm contestation and negotiation wherein [...] positions and [...] practices are constructed and performed."[100]

Despite the wide acknowledgement that training can bring about positive change, providing consistent training to staff deployed to peace operations still faces a range of obstacles. Although training is ever evolving, there are still gaps, for example the failure to tackle issues such as gender-based violence and the prevention of sexual exploitation and abuse through a gender lens.[101] In addition, gender training is not systematically provided and differs across contributing states.[102] Training appears to have been more consistent in Western countries, where there are more training opportunities, when compared to the Global South.[103] This results in peace operations that consist of experts that have had widely varying exposure to gender training, which inevitably will have an impact on the likelihood of change.

While many training manuals are openly available online, and continuous revisions are moving away from a purely top-down institutional focus on gender training, little is known about the impact of different ways of introducing gender in pre-deployment training. To ensure positive change, it is vital to better understand the effectiveness and impact of gender training through more enhanced and transparent evaluation.[104] This will feed into lessons learned on gender mainstreaming and will allow for measuring change more systematically.[105]

---

99 Aiko Holvikivi, "Training the troops on gender: the making of a transnational practice," *International Peacekeeping* 28, no. 2 (2021): 175-199.

100 Georgina Holmes, "Situating agency, embodied practices and norm implementation in peacekeeping training," *International Peacekeeping* 26, no. 1 (2019): 55-84, 55.

101 Kristin Kvinnesland, "Peacekeeper Training on the Prevention of Sexual Exploitation and Abuse: Adding a gendered perspective," Master's thesis, UiT Norges arktiske universitet, 2021.

102 Carson, "Pre-deployment 'gender' training"; Comfort Lamptey, "Gender Training in United Nations Peacekeeping Operations," *Gender and Peacekeeping in Africa, Occasional Series* 5 (2012); Minna Lyytikainen, "Gender Training for Peacekeepers: *Preliminary Overview of United* Nations peace support operations *United Nations International Research and Training Institute for the Advancement of Women (UN-INSTRAW)* (2007).

103 Lyytikainen, "Gender Training for Peacekeepers".

104 Lyytikainen, "Gender Training for Peacekeepers".

105 Ferguson, "Theory of Change".

Conclusion

Scholars exploring the implementation of gender mainstreaming in operational environments have focused on the attitude of deployed experts towards gender issues, the mainstreaming of gender in mission structures, and the role of gender mainstreaming in the implementation of the mission mandate. This chapter has engaged with some of the key issues and debates that are prevalent in the field of gender mainstreaming in peace operations. It has covered institutional and policy change, as well as ongoing conceptual discussions, focusing on three key debates: the conflation of 'gender' and 'woman', 'add women and stir', and the need for decolonial thinking. It has demonstrated that conceptual challenges not only impact academic debates, but also the implementation of gender mainstreaming in peace operations in conflict-affected environments. It has finally discussed the importance of training for the successful implementation of gender mainstreaming.

From the literature reviewed, much has been done since the concept of gender mainstreaming was first introduced into the institutional architecture of peace operations. While this is something to be celebrated, several challenges persist. Feminist research has pointed at the co-optation[106] of gender mainstreaming into existing power structures and reluctance to adopt intersectional and decolonial approaches. Further research on in-depth case studies and cross-disciplinary cooperation is needed to enhance our understanding of gender mainstreaming, and to further bridge the gap between academic knowledge and successful implementation in peace operations.

References

Ackerly, Brooke and True, Jacqui. 2008. "Reflexivity in practice: Power and ethics in feminist research on international relations." *International Studies Review*, 10(4), p693-707.

Anderlini, Sanam Naraghi. "Civil society's leadership in adopting 1325 resolution." In *The Oxford Handbook of Women, Peace, and Security*, edited by Sara E. Davies and Jacqui True. Oxford: Oxford University Press, 2018.

106 Sara De Jong and Susanne Kimm. "The co-optation of feminisms: A research agenda." *International Feminist Journal of Politics* 19, no. 2 (2017): 185-200.

Ansorg, Nadine, and Toni Haastrup. "Gender and the EU's Support for Security Sector Reform in Fragile Contexts." *JCMS: journal of common market studies* 56, no. 5 (2018): 1127-1143.

Bargués-Pedreny, Pol, and Maria Martin de Almagro. "Prevention from Afar: gendering resilience and sustaining hope in post-UNMIL Liberia." *Journal of Intervention and Statebuilding* 14, no. 3 (2020): 327-348.

Basini, Helen SA. "Gender mainstreaming unraveled: The case of DDRR in Liberia." *International Interactions* 39, no. 4 (2013): 535-557.

Basu, Soumita, Paul Kirby, and Laura J. Shepherd. "Women, Peace and Security: a critical cartography." in *New Directions in Women, Peace and Security*, edited by Soumita Basu, Paul Kirby, and Laura J. Shepherd. Bristol, Bristol University Press, 2020.

Beber, Bernd, Michael J. Gilligan, Jenny Guardado, and Sabrina Karim. "Peacekeeping, compliance with international norms, and transactional sex in Monrovia, Liberia." *International Organization* 71, no. 1 (2017): 1-30.

Bridges, Donna, and Debbie Horsfall. "Increasing operational effectiveness in UN peacekeeping: Toward a gender-balanced force." *Armed Forces & Society* 36, no. 1 (2009): 120-130.

Capan, Zeynep Gulsah, Filipe Dos Reis, and Maj Grasten. *The Politics of Translation in International Relations*, Palgrave Macmillan, Cham, 2021.

Carreiras, Helena. "Gendered culture in peacekeeping operations." *International Peacekeeping* 17, no. 4 (2010): 471-485.

Carson, Lisa. "Pre-deployment 'gender'training and the lack thereof for Australian peacekeepers." *Australian Journal of International Affairs* 70, no. 3 (2016): 275-292.

Charlesworth, Hilary. "Not waving but Drowning: Gender Mainstreaming and Human Rights in the United Nations." *18 HARV. Hum. Rts.* J 1 (2005).

Cohn, Carol, ed. *Women and wars: Contested histories, uncertain futures.* John Wiley & Sons, 2013.

Daly, Mary. "Gender mainstreaming in theory and practice." *Social Politics: International Studies in Gender, State & Society* 12, no. 3 (2005): 433-450.

Davis, Lisa, and Jessica Stern. "WPS and LGBTI rights." in *The Oxford Handbook of Women, Peace, and Security*, edited by Sara E. Davies and Jacqui True. Oxford: Oxford University Press, 2018.

De Jong, Sara, and Susanne Kimm. "The co-optation of feminisms: A research agenda." *International Feminist Journal of Politics* 19, no. 2 (2017): 185-200.

DeGroot, Gerard J. "A few good women: Gender stereotypes, the military and peacekeeping." *International Peacekeeping* 8, no. 2 (2001): 23-38.

Deiana, Maria-Adriana, and Kenneth McDonagh. "'It is important, but…': translating the Women Peace and Security (WPS) Agenda into the planning of EU peacekeeping missions." *Peacebuilding* 6, no. 1 (2018): 34-48.

Deiana, Maria-Adriana, and Kenneth McDonagh. "Translating the women, peace and security agenda into EU common security and defence policy: Reflections from EU peacebuilding." *Global Society* 32, no. 4 (2018): 415-435.

Dharmapuri, Sahana "Why is Gender Equality in Peacekeeping Important" in *Not Just a Numbers Game. Increasing Women's Participation in UN Peacekeeping* (International Peace Institute, 2013).

Duncanson, Claire. "Forces for good? Narratives of military masculinity in peacekeeping operations." *International Feminist Journal of Politics* 11, no. 1 (2009): 63-80.

Duncanson, Claire. *Forces for good?: Military masculinities and peacebuilding in Afghanistan and Iraq*. Springer, 2013.

Duriesmith, David. *Masculinity and New War: The gendered dynamics of contemporary armed conflict*. Routledge, 2016.

Egnell, Robert, Petter Hojem, and Hannes Berts. *Gender, military effectiveness, and organizational change: the Swedish model*. Berlin: Springer, 2014.

Enloe, Cynthia. "Being Reflexively Feminist Shouldn't Be Easy." In *Researching War. Feminist Methods, Ethics and Politics,* edited by Annick T. R. Wibben. London: Routledge, 2016.

Fara-Zamora, Jorge L, Andrés Molina-Ochoa, Nancy C. Doubleday, eds., *The Colombian Peace Agreement. A Multidisciplinary Assessment*. Routledge, New York, 2021.

Fernández, Marta, and Lucas Guerra. "Peacebuilding and Postcolonial Subject." In *The Palgrave Encyclopedia of Peace and Conflict Studies,* edited by Oliver Richmond and Gezim Visoka, London: Palgrave Macmillan, 2020.

Galtung, Johan, and Dietrich Fischer. "Positive and Negative Peace." In *Springer Briefs on Pioneers in Science and Practice*, ed. Johan Galtung, vol 5. Springer, Berlin.

Gibbings, Sheri Lynn. "No angry women at the United Nations: political dreams and the cultural politics of United Nations Security Council Resolution 1325." *International Feminist Journal of Politics* 13, no. 4 (2011): 522-538.

Guerrina, Roberta, and Katharine AM Wright. "Gendering normative power Europe: lessons of the Women, Peace and Security agenda." *International affairs* 92, no. 2 (2016): 293-312.

Haastrup, T. and Jamie J. Hagen, "Global racial hierarchies and the limits of localization via National Action Plans", in *New Directions in Women, Peace and Security,* edited by Soumita Basu, Paul Kirby, and Laura J. Shepherd, Bristol, Bristol University Press, 2020.

Hagen, Jamie J. "Queering women, peace and security." *International Affairs* 92, no. 2 (2016): 313-332.

Hall, Stuart. *Essential essays, volume 1: Foundations of cultural studies*. Duke University Press, 2018.

Harrington, Carol. "Resolution 1325 and post-Cold War feminist politics", *International Feminist Journal of Politics*, 13, no. 4 (2011): 557-575.

Heinecken, Lindy. "Are Women 'Really' Making a Unique Contribution to Peacekeeping?: The Rhetoric and the Reality." *Journal of International Peacekeeping* 19, no. 3-4 (2015): 227-248.

Holmes, Georgina. "Situating agency, embodied practices and norm implementation in peacekeeping training." *International Peacekeeping* 26, no. 1 (2019): 55-84.

Holvikivi, Aiko. "Training the troops on gender: the making of a transnational practice." *International Peacekeeping* 28, no. 2 (2021): 175-199.

Hord, Shirley M., and James L. Roussin. *Implementing change through learning: Concerns-based concepts, tools, and strategies for guiding change*. Corwin Press, 2013.

"HQ SACT Office of the Gender Adviser", NATO, https://www.act.nato.int/gender-advisor

Huber, Laura K., and Natalie F. Hudson. "Deepening the Conversation: Feminism, International Policing and the WPS Agenda." *International Peacekeeping* 26, no. 5 (2019): 579-604.

Hudson, Heidi. "'Doing'security as though humans matter: A feminist perspective on gender and the politics of human security." *Security Dialogue* 36, no. 2 (2005): 155-174.

Hudson, Heidi. "Decolonising gender and peacebuilding: feminist frontiers and border thinking in Africa." *Peacebuilding* 4, no. 2 (2016): 194-209.

Hudson, Heidi. "Decolonizing the mainstreaming of gender in peacebuilding: toward an agenda for Africa." African Peacebuilding Network (APN) Working Paper 8 (2016).

Hudson, Heidi. "It Matters How You 'Do'Gender in Peacebuilding: African Approaches and Challenges." *Insight on Africa* 13, no. 2 (2021): 142-159.

Hudson, Natalie Florea. "En-gendering UN peacekeeping operations." *International Journal* 60, no. 3 (2005): 785-807.

Jacobs, An D., and Katerina Krulisova. "Gender Mainstreaming in Peacebuilding." *LSE Women Peace and Security Policy Brief Series*, no. 4, 2021. https://www.lse.ac.uk/women-peace-security/assets/documents/2021/Gender-Mainstreaming-in-Peacebuilding-Policy-Brief-04.pdf

Jenne, Nicole, and Fiorella Ulloa Bisshopp. "Female Peacekeepers: UNSC Resolution 1325 and the Persistence of Gender Stereotypes in the Chilean Armed Forces." *International Peacekeeping* 28, no. 1 (2021): 134-159.

Jennings, Kathleen M. "Unintended consequences of intimacy: political economies of peacekeeping and sex tourism." *International Peacekeeping* 17, no. 2 (2010): 229-243.

Jennings, Kathleen M., and Morten Bøås. "Transactions and interactions: Everyday life in the peacekeeping economy." *Journal of Intervention and Statebuilding* 9, no. 3 (2015): 281-295.

Karamé, Kari H. "Military women in peace operations: Experiences of the Norwegian battalion in UNIFIL 1978–98." *International Peacekeeping* 8, no. 2 (2001): 85-96.

Karim, Sabrina and Marsha Henry. "Gender and peacekeeping" in *The Oxford Handbook of Gender and Conflict*, eds. Fionnuala Ní Aoláin, Naomi Cahn, Dina Francesca Haynes and Nahla Valji. Oxford: Oxford University Press, 2018.

Karim, Sabrina, and Kyle Beardsley. "Explaining sexual exploitation and abuse in peacekeeping missions: The role of female peacekeepers and gender equality in contributing countries." *Journal of Peace Research* 53, no. 1 (2016): 100-115.

Karim, Sabrina, and Kyle Beardsley. "Female peacekeepers and gender balancing: Token gestures or informed policymaking?." *International Interactions* 39, no. 4 (2013): 461-488.

Kathleen Jennings, 'Protecting Whom? Approaches to Sexual Exploitation and Abuse in UN Peacekeeping Operations', Oslo: FAFO Report, 2008.

Kreft, Anne-Kathrin. "The gender mainstreaming gap: Security Council resolution 1325 and UN peacekeeping mandates." *International Peacekeeping* 24, no. 1 (2017): 132-158.

Kronsell, Annica. "Sexed bodies and military masculinities: Gender path dependence in EU's common security and defense policy." *Men and Masculinities* 19, no. 3 (2016): 311-336.

Krook, Mona Lena, and Jacqui True. "Rethinking the life cycles of international norms: The United Nations and the global promotion of gender equality." *European journal of international* relations 18, no. 1 (2012): 103-127.

Kunz, Rahel, Elisabeth Prügl, and Hayley Thompson. "Gender expertise in global governance: Contesting the boundaries of a field." *European Journal of Politics and Gender* 2, no. 1 (2019): 23-40.

Kunz, Rahel. "Messy feminist knowledge politics: a double reading of post-conflict gender mainstreaming in Liberia." *International Feminist Journal of Politics* 22, no. 1 (2020): 63-85.

Kvinnesland, Kristin. "Peacekeeper Training on the Prevention of Sexual Exploitation and Abuse: Adding a gendered perspective." Master's thesis, UiT Norges arktiske universitet, 2021.

Lahoud, Nina J. "What Fueled the Far-Reaching Impact of the Windhoek Declaration and Namibia Plan of Action as a Milestone for Gender Mainstreaming in UN Peace Support Operations and Where Is Implementation 20 Years Later?." *Journal of International Peacekeeping* 24, no. 1-2 (2020): 1-52.

Lamptey, Comfort. "Gender Training in United Nations Peacekeeping Operations." *Gender and Peacekeeping in Africa, Occasional Series* 5 (2012).

Landgren, Karin. "WPS and Gender Mainstreaming." In *The Oxford Handbook of Women, Peace, and Security,* edited by Sara E. Davies and Jacqui True. Oxford: Oxford University Press, 2018.

Lyytikainen, Minna. "Gender Training for Peacekeepers: *Preliminary Overview of United* Nations peace support operations *United Nations International Research and Training Institute for the Advancement of Women (UN-INSTRAW)* (200.7)

Mackay, Fiona, and Georgina Waylen. "Introduction: Gendering "new" institutions." *Politics & Gender* 10, no. 4 (2014): 489-494.

Martín de Almagro, María. "Producing participants: Gender, race, class, and women, peace and security." *Global Society* 32, no. 4 (2018): 395-414.

McDonagh, Kenneth and Maria-Adriana Deiana. "Add women and hope? Assessing the gender impact of EU common security and defence policy missions: a policy report." (Dublin: Dublin City University, 2017).

Messerschmidt, Maike, and Hendrik Quest. "Peacebuilding measures and the transformation of masculinities: Looking at Liberia and Uganda." In *Gender Roles in Peace and Security*, 79-100. Springer, Cham, 2020.

Ministry of Defence. "JSP 985 Human Security in Defence Volume 1: Incorporating Human Security in the way we Operate." 2021. Available online at: https://assets.publishing.service.gov.uk/government/uploads/system/uploads/attachment_data/file/1040257/20211209_JSP_985_Vol_1.pdf

Moncrief, Stephen. "Military socialization, disciplinary culture, and sexual violence in UN peacekeeping operations." Journal of Peace Research 54, no. 5 (2017): 715-730.

NATO. "NATO/EAPC Women, Peace and Security Policy and Action Plan 2018." 2018. Available online at: https://www.nato.int/nato_static_fl2014/assets/pdf/pdf_2018_09/180920-WPS-Action-Plan-2018.pdf

Nduka-Agwu, Adibeli. "'Doing gender'after the war: Dealing with gender mainstreaming and sexual exploitation and abuse in UN peace support operations in Liberia and Sierra Leone." *Civil Wars* 11, no. 2 (2009): 179-199.

Newby, Vanessa F., and Clotilde Sebag. "Gender sidestreaming? Analysing gender mainstreaming in national militaries and international peacekeeping." *European Journal of International Security* 6, no. 2 (2021): 148-170.

Olsson, Louise, and Theodora-Ismene Gizelis. "Advancing gender and peacekeeping research." *International Peacekeeping* 21, no. 4 (2014): 520-528.

Olsson, Louise. "Mainstreaming gender in multidimensional peacekeeping: A field perspective." *International Peacekeeping* 7, no. 3 (2000): 1-16.

Olsson, Louise. *Gender equality and United Nations peace operations in Timor Leste* (Leiden: Martinus Nijhoff Publishers, 2009).

Parashar, Swati. "The WPS agenda." In *The Oxford Handbook of Women, Peace, and Security,* edited by Sara E. Davies and Jacqui True. Oxford: Oxford University Press, 2018.

Peacewomen.org. "1325 National Action Plans (NAPs)." Available online at: http://1325naps.peacewomen.org/.

Pruitt, Lesley J. "All-female police contingents: Feminism and the discourse of armed protection." *International Peacekeeping* 20, no. 1 (2013): 67-79.

Reeves, Audrey. "Feminist knowledge and emerging governmentality in UN peacekeeping: Patterns of co-optation and empowerment." *International Feminist Journal of Politics* 14, no. 3 (2012): 348-369.

Richmond, Oliver P., and Roger Mac Ginty. "Where now for the critique of the liberal peace?." *Cooperation and Conflict* 50, no. 2 (2015): 171-189.

Ruiz, Andrea Luna. "Gender Approach: A Requirement for Peacebuiding." *Global Journals of Human-Social Science Linguistics and Education* 19, no. 1 (2019): 23-30.

Rupesinghe, Natasja, Eli Stamnes, and John Karlsrud. "WPS and female peacekeepers." In *The Oxford Handbook of Women, Peace, and Security,* edited by Sara E. Davies and Jacqui True. Oxford: Oxford University Press, 2018.

Sabaratnam, Meera. "Avatars of Eurocentrism in the Critique of the Liberal Peace." *Security Dialogue* 44, no. 3 (2013): 259-278.

Shepherd, Laura J. "Victims of violence or agents of change? Representations of women in UN peacebuilding discourse." *Peacebuilding* 4, no. 2 (2016): 121-135.

Shepherd, Laura J. Gender, UN Peacebuilding, and the Politics of Space: Locating Legitimacy. Oxford: Oxford University Press, 2017.

Simić, Olivera. "Does the presence of women really matter? Towards combating male sexual violence in peacekeeping operations." *International Peacekeeping* 17, no. 2 (2010): 188-199.

Smith, Sarah. "Accountability and sexual exploitation and abuse in peace operations." *Australian Journal of International Affairs* 71, no. 4 (2017): 405-422.

Towns, Ann. "The status of women as a standard of 'civilization'." *European Journal of International Relations* 15, no. 4 (2009): 681-706.

True, Jacqui. "How effective is gender mainstreaming in international peace and security policymaking?." In Handbook on Gender in World Politics. Edward Elgar Publishing, 2016.

True, Jacqui. "How effective is gender mainstreaming in international peace and security policymaking?." In *Handbook on Gender in World Politics*. Edward Elgar Publishing, 2016.

True, Jacqui. "The Unfulfilled Mandate-Gender Mainstreaming and UN Peace Operations." *Georgetown Journal of International Affairs* 10, Issue 2 (2009): 41-50.

UNICEF – United Nations International Children's Emergency Fund, 2016. Learning for peace. Available at: https://eccnetwork.net/wp-content/uploads/Gender-Education-and-Peacebuilding-A-Review-of-Learning for-Peace-Case-Studies.pdf [Accessed 18 August 2021].

United Nations, Department of Peacekeeping Operations, Department of Field Support. "Gender Responsive United Nations Peacekeeping Operations." Available online at: https://peacekeeping.un.org/sites/default/files/english_gender_responsive_united_nations_peacekeeping_operations_policy_1.pdf.

United Nations. "ECOSOC Agreed Conclusions 1997/2." 1997. Available online at: https://www.un.org/womenwatch/osagi/pdf/ECOSOCAC1997.2.PDF

Valenius, Johanna. "A few kind women: gender essentialism and Nordic peacekeeping operations." *International peacekeeping* 14, no. 4 (2007): 510-523.

Vlachová, Marie, and Lea Biason, eds. Women in an insecure world: violence against women: facts, figures and analysis. Dcaf, 2005.

Westendorf, Jasmine-Kim, and Louise Searle. "Sexual exploitation and abuse in peace operations: trends, policy responses and future directions." *International Affairs* 93, no. 2 (2017): 365-387.

Westendorf, Jasmine-Kim. "'Add women and stir': the Regional Assistance Mission to Solomon Islands and Australia's implementation of United Nations Security Council Resolution 1325." *Australian Journal of International Affairs* 67, no. 4 (2013): 456-474.

Whitworth, Sandra. *Men, militarism, and UN peacekeeping: A gendered analysis.* Lynne Rienner Publishers, 2004.

Wilén, Nina. "Female peacekeepers' added burden." *International Affairs* 96, no. 6 (2020): 1585-1602.

WPS Focal Points Network. "Regional Action Plans." Available online at: https://wpsfocalpointsnetwork.org/regional-action-plans/

"What is a Peace Operation", Center for International Peace Operations (ZIF), accessed 30 May 2022, https://www.zif-berlin.org/en/what-peace-operation)

2

US MILITARY GOVERNMENT, CIVIL AFFAIRS, AND WPS

Expanding Participation Through Partnership

Spencer B. Meredith III[1]

In 2017, the United States established the *Women, Peace and Security (WPS) Act*, followed two years later by the *US Strategy on Women, Peace, and Security*. Even though the formalization of the WPS principles has been recent, the acknowledgement and prioritization of women as vital participants has a longer history in the US. This is in part due to the democratic nature of the Republic and the expansion of its political franchise; a key evolution of the US historical approach to WPS principles has been increasing women's *participation* in governance – the processes of decision making and implementation within various types of political, economic, and social systems. Other WPS priorities of *prevention* and *protection* have occurred more implicitly in efforts to prevent and manage conflict through diplomatic and development assistance, as well as military government activities designed to maintain distinctions of non-combatants and post-conflict rule of law protections.[2] Consequently, while not exclusively focused on women's participation, military government has created favorable contexts for

1 Views expressed do not represent official US government or National Defense University positions.

2 According to US Department of Defense Directive 2000.13, military government occurs under conditions following the collapse of civilian governments in foreign countries, or as US forces gain control over territory after combat operations. As a result, it can either support existing civilian administrations or utilize transitional military authority prior to the establishment of legitimate civilian control.

participation through the protection of civilian populations and support for local communities. This has always been paramount to successful military operations and consistent with the WPS pillars. This chapter focuses on historical efforts to protect populations and increase women's participation in governance through the use of military government broadly, and Civil Affairs operations more specifically.[3] It also highlights the evolution of women's participation within the US military and looks to this evolution as a primary focus for building capacity through the increased participation of local women in partner states.

Military Support to Governance and Women's Participation

The United States has a long history of using military forces to govern. Military government has included post-conflict stabilization of newly acquired territories, as well as supporting transitions to permanent civilian administrations. Generally, the military has been a temporary agent of government, often following the cessation of combat operations in formerly hostile territory. As a result, US military government approaches have combined elements of counterinsurgency with the establishment of effective public infrastructures to support local populations. These activities have required US military government forces to understand and utilize local governance before eventually turning over their responsibilities to civilian authorities.

In a broad sense, governance processes occur at all levels of communal organizations. They can be formal, codified regulations as much as informal 'word of mouth' norms passed down generationally. From family and clans, to corporations and international institutions, governance is the convergence of power and legitimacy. Accordingly, the US military approach has emphasized participation and engagement with population *centers of gravity* as legitimate sources of power. These provide moral or physical strength, freedom of action, and the will to act.[4] In particular, US Army Civil Affairs units focus on human networks as the mechanism by

3 The proper noun 'Civil Affairs' used throughout this chapter refers to US military units conducting governance-related operations within partner countries or during post-conflict stabilization efforts. It is commonly used within the US military to denote the operations themselves, as well as the personnel, training, tasks, and authorities to conduct the same.

4 US Joint Publication 1-02 Department of Defense Dictionary of Military and Associated Terms (2008): 81.

which individual and communal interests can be articulated, aggregated, and acted upon in governance systems.[5] This 'human domain' has been codified in various US military documents, but more importantly socialized in a culture of 'people-centered' military operations within US Civil Affairs.[6] Given that woman play central roles across human networks, even if not universally equitably, effective military governance and Civil Affairs operations require that women participate fully as agents in the process.[7]

The US *Women, Peace, and Security Congressional Report* (2022) details developments in the Departments of State, Defense, Homeland Security, and the Agency for International Development across the WPS spectrum. Initiated under the *Women, Peace, and Security Act* of 2017, the Report indicates increased analysis of gender within US bureaucracies and partner assistance programs. Specifics include support to over five million gender-based violence survivors globally, work with women leaders to assist with refugee resettlement efforts in post-conflict areas, and increased participation of women in US and partner military and police forces.[8] A central purpose of those efforts is to increase legitimacy of effective governance through increased participation of women.[9]

Despite the relatively new focus on WPS priorities as discrete military requirements in US law and doctrine, the principles they embody have a long history in US military governance, both through the development of a more balanced force at home, and through increased efforts to support women as key security partners abroad. Building on the legacy of female combat nurses during the Revolutionary and Civil Wars, through the Women's Army Corps and civilian employment in critical industries during World War II, to contemporary military Female Engagement Teams, the goals of the WPS have advanced in all areas of the US military.

---

5 US Army Field Manual 3-0 *Operations* (2017); US Army Field Manual 3-57 *Civil Affairs* (2021). This chapter discusses US military Civil Affairs broadly, recognizing that the preponderance of capabilities resides in the Army.

6 US Department of Defense, *Summary of the Irregular Warfare Annex to the National Security Strategy* (2020); US Joint Publication 3-13, *Information Operations* (2014); US Army Field Manual 3-07 *Stability* (2014); US Army Field Manual 3-24, *Insurgencies and Countering Insurgencies* (2014).

7 For a concise review of relevant scholarship, see Valerie Hudson *et al.*, 'Clan Governance and State Stability: The Relationship between Female Subordination and Political Order,' *American Political Science Review*, 109, no.3, (2015): 535-555.

8 US Government Women, Peace, and Security Congressional Report (July 2022), 5.

9 One illustrative example cited in the report highlighted working with local groups in Ghana to include more than 30,000 women in agricultural governance programs ranging from advocacy training to technical support. US Government *Women, Peace, and Security Congressional Report* (July 2022), 25.

In particular, as part of the broader Department of Defense's *Women, Peace, and Security Framework and Implementation Plan* (2020), and building on a US Army TRADOC *Gender Integration Study* (2015), Civil Affairs consider the role of women in several ways to include recruitment and training of US personnel, mission tasks, and external engagement with partners.[10]

Much of the explicit focus on women's participation occurred after the establishment of women as a permanent part of the US military following their pervasive role during World War II. Increasing participation of women in the US military has broadened especially over the past 30 years, to include piloting fighter aircraft, commanding tanks, and graduating from elite infantry schools. As the lead military entity tasked with engaging civilian populations in support of civilian governance, Civil Affairs have also focused on women's participation as a force provider, and in support of local communities and their governments.[11]

The most recent US Army Field Manual on military governance (3-57, 2021) describes Civil Affairs operations as engagement with and leverage of key civil elements within an operational environment, to include women stakeholders in political, economic, and social arenas. Civil Affairs core tasks include 1) post-conflict transitional governance in support of stabilization efforts, 2) civil network engagement and development, and 3) civil-military coordination across US and partner entities, spanning local security and infrastructure management, to rule of law and policy considerations.[12] Consequently, Civil Affairs soldiers are required to possess general knowledge of governance, while also specializing in either functional areas (for example, engineering, education, facilities management) or operational environments defined along a continuum of *cooperation—competition below armed conflict—conflict*, both of which require engagement with all

10 Following the 2015 US Army Secretary Directive 2015-08 to open more than 4,000 positions in the US Army Special Operations Command, some examples include the *Army People Strategy – Diversity, Equity and Inclusion Annex* (2020) integrating gender equality as part of training, education, and selection outcomes; 'Project Inclusion' initiated to identify and address issues hindering equal consideration during promotion boards (for example, removing names and photos); increasing opportunities in combat arms specializations (for example, the first gender integrated Infantry One Station Unit Training in 2017); and improving holistic training for women to increase pass rates on Army Combat Fitness Test. Remarks to the US Senate Armed Services Committee (April 2022) by the US Army Special Operations Commander highlighted a Women in ARSOF (Army Special Operations Forces) Study for many of those advances.

11 From the earliest days of Operation Enduring Freedom in Afghanistan, women soldiers had 'boots on the ground,' many of them in Civil Affairs roles. Those roles expanded as US policies shifted from unconventional warfare to remove the Taliban and al Qaeda safe havens, towards state and nation building across the country.

12 US Army Field Manual 3-57 *Civil Affairs Operations* (July 2021), 1-1.

parts of populations, men and women, and their increasing participation in governance.[13]

Historical Evolutions

The emphasis on engagement and participation is part of a longer tradition within US military government operations that began after the invasion of colonial Canada failed to gain the support of the local population.[14] Following the establishment of the United States, westward expansion brought into sharp focus the need to reduce frontier hostilities between communities. Yet despite a growing body of treaties with indigenous groups, the guiding principles of US military government remained limited to protecting settlers and putting down rebellions.[15] More often than not, soldiers guarded the peace and protected settlers against tribal communities, rather than alongside them.

However, two notable exceptions shaped US military government approaches: the US-Mexican War (1846-1848) and the acquisition of the Alaska Territory (1867). US military government of Mexico formalized distinctions between unconquered and conquered territories, producing different forms and functions for garrisoning US troops.[16] It also emphasized protecting civilians and their properties, fostering local markets, maintaining law and order (and the prevention and punishment of looting), and restoration of local legal authorities. Identifying and supporting the participation of local stakeholders was key to the success of those operations, the goal of which was to restore local civilian governance. Similar attention to preserving community governance under US military government occurred in the Alaskan territory in 1867. Following a mostly 'hands-off' approach due to the lack of instigating friction between the few

---

13 US Joint Doctrine Note 1-19 Competition Continuum (June 2019), 3; US Joint Publication 1-02 Joint Operating Concept—Irregular Warfare: Countering Irregular Threats (May 2010), 9.
14 US Army Field Manual 41-5, *Joint Manual for Civil Affairs* (1966) illustrates many of the common goals and tasks between the current and previous post-World War II/Vietnam era approaches. The shift towards greater partner nation involvement today is telling given the increased competition for influence in the current system. See Nadia Shadlow, *War and the Art of Governance: Consolidating Combat Success to Political Victory* (Washington: Georgetown University Press, 2017) for an effective overview.
15 See Michael Nunnally, American Indian Wars: a Chronology of Confrontations between Native Peoples and Settlers and the United States Military (London: MacFarland and Co., 2007).
16 'The Mexican War and After,' *American Military History* (Washington: US Army Center of Military History, 1989), 163-183.

American settlers and local communities, military government was often limited to maintaining peaceful relations between local tribes. Violent encounters occasionally occurred, usually arising from localized conflicts that escalated to community-wide threats to US control. These were sometimes met with overwhelming force, but the general trend in Alaska leaned more towards peaceful conflict resolution through non-interference in local affairs, while maintaining general stability to enable them.[17]

US intervention in the Russian Far East in 1918 also emphasized preserving local governance and protection of civilian populations as key parts of its military presence. While the deployment of nearly 10,000 American Expeditionary Forces to Vladivostok was initially envisioned to protect military supplies and soldiers committed to the Alliance, US soldiers soon found themselves embroiled with their allies' efforts to advance geopolitical aims there, in particular Japanese predations in the region. Officially tasked with not inhibiting local political activities in a tumultuous political climate, US forces also contended with attacks on civilians by Tsarist forces, and later engaged in combat with Bolshevik troops targeting the same.[18] Despite the risks of escalating international tensions, US troops focused on protecting political freedoms and public safety as primary concerns over the two year deployment, similarly to military government operations in the Rhineland during the same period.

Compared to previous occupations in Mexico, the North American plains, Cuba and the Philippines, the US presence in Germany's industrial heartland (1918-1920) necessitated even greater coordination with local governance to ensure local stability.[19] However, the effort was often constrained by increased tension between allied forces. Whereas Alaska had been relatively 'hands-off' in the day-to-day activities of local communities, the Rhineland was directly 'hands-on' working in partnership with local government officials and community leaders. In particular, US Army forces maintained and utilized existing courts, police, laws, and civil administrations – all key stakeholders in the governance of villages and cities under foreign occupation. The only restrictions on German governance activities were to prevent interference with American troops fulfilling

17 Stanley Sandler, Glad to See Them Come and Sorry to see them go: A History of US Army Tactical Civil Affairs/Military Government, 1775-1991 (1994) provides an excellent analysis of primary source documents for both cases.

18 William Graves, *America's Siberian Adventure: 1918-1920* (American History Classics, 2019).

19 American Military Government of Occupied Germany: 1918-1920 (Washington: US Army, 1943)

their duties or undermining their security. Thus, while the presence of US troops remained small compared to other allied militaries in the region, the generally positive reception by local populations was instructive for later military government operations.[20] Most notably it emphasized local participation and partnership with the population, rather than control over it.

World War II marked a turning point in the evolution of US military government and its emphasis on local participation. In particular, de-Nazification sought to remove active agents of the former regimes, while drawing distinctions within the civilian and military populations to ensure stable local governance. Recognizing that many were directly compelled to join the fascist system, US standards distinguished willing participation from compulsory service to the government. This nuanced approach developed during specialized governance training for civilian experts who volunteered to form Office of Military Government, US (OMGUS) personnel.[21] OMGUS sought to mobilize the German population and surviving military and government members as key participants in reconstruction efforts, while also assisting the prosecution of war crimes.

World War II also saw an increase in the roles and number of US women deployed overseas in support of those efforts. Of the nearly 150,000 soldiers in the US Women's Army Corps, several thousand remained in occupied Germany to assist with US Army reconstruction.[22] Some of the critical functions included family reunification, reestablishment of local constabulary units, and food supplies.[23] As a result, the increasing participation of women within the US military and as an emphasis for post-conflict stabilization, contributed to the expansion of both elements decades later during the conflict in Vietnam, effectively laying the foundation for later WPS priorities.[24]

---

20 Sandler, 146-8.

21 Susan Carruthers, *The Good Occupation: American Soldiers and the Hazards of Peace* (Cambridge: Harvard University Press, 2016) examines the role of civilian professionals trained to support military government as functional specialists in everything from engineering and construction, to banking and agriculture.

22 Mattie Treadwell, *The Women's Army Corps* (Washington: US Army, 1991) provides a comprehensive treatment of the WAC.

23 See Thomas Maulucci, and Detlef Junker, GIs in Germany: The Social, Economic, Cultural, and Political History of the American Military Presence (Washington: Cambridge University Press, 2013).

24 US Civil Affairs doctrine was greatly clarified during that time, building primarily on the legacies of World War II to define mission requirements and goals for partner participation. Women soldiers and military engagement with local communities of women remained key elements of US counterinsurgency and regional stabilization efforts.

However, US military government in Iraq beginning in 2003 presented a more 'Janus-faced' approach to governance and women's participation than its immediate predecessors. Unlike the earlier, more nuanced de-Nazification of post-World War II Germany, de-Baathification of Iraq widely cast out members of the Ba'ath party at all levels of involvement, even as US military government greatly increased the role of women soldiers and support to Iraqi women. In doing so, the US sought to empower key participants in Iraqi society, while effectively neutralizing the only indigenous partner force capable of working to restore and maintain public safety for them.[25] As large segments of the Iraqi military, police force, and civil service were forced out of their jobs, many took up arms as participants in the Sons of Iraq insurgency.[26] Meanwhile, US efforts worked to expand the participation of women across Iraqi governance systems, often creating conflicting imperatives between protection and participation for many Iraqi civilians.

Despite the profound policy disjuncture, some areas saw success in partnering with local populations. Most notable were women's participation in government, business development, and advocacy groups in different regions. Variations in the scale of success occurred depending on the level of violence amidst inter-community conflict, but overall progress was made.[27] In particular, women's participation in public sector employment and small business development have continued to expand after the withdrawal of US Forces in 2011.[28]

The participatory model used by US Civil Affairs to support this empowerment of women bears mentioning, in particular the method of integrating business leaders, community caregivers, and family organizations to support local populations. For example, in 2004, members of the 422nd Civil Affairs Battalion started partnering with the Director of the Dina Institute to expand her care for mentally and physically disabled children in the capital. Supported by a microfinance loan of approximately

---

25 Peter Stirk, *The Politics of Military Occupation* (Edinburgh: Edinburgh University Press, 2009) details many of these complexities. See also Andrew Rathmell, 'Planning Post-Conflict Reconstruction in Iraq: What Can We Learn,' *International Affairs*, 81, no.5 (2005): 1013-38.

26 Eric Henning and Glen Rangwala, *Iraq in Fragments* (London: Hurst, 2006) identify many of the failures of US military governance of Iraq.

27 Luca Bandiera *et al.* Jobs in Iraq: A primer on Job Creation in the Short-Term (World Bank, 2019).

28 Doaa Althalathini *et al.* 'The Impact of Islamic Feminism in Empowering Women's Entrepreneurship in Conflict Zones: Evidence from Afghanistan, Iraq, and Palestine' *Journal of Business Ethics*, 178, no.1 (2022): 39-55.

$5000 to the Institute, the Civil Affairs team also connected the Director with one of the largest construction firms in Iraq and a women-led local NGO providing medical care and aid to families in Baghdad. Highlighting shared values regarding the role of women as critical agents of community wellbeing, the Civil Affairs team enabled longer-term financial and material support for the Dina Institute by expanding network connections between otherwise disassociated groups. Over the course of several months, the Civil Affairs team facilitated fuel deliveries, new clothing for the children living at the institute, and improvements to water and electricity systems. The team also worked closely with local police and government officials to support the facility's continued operation. While the Dina Institute was a comparatively small project within the larger US reconstruction effort, it nevertheless presents an example of prioritizing local networks in support of women's participation.[29]

US Civil Affairs in Afghanistan expanded that focus on women as critical social network stakeholders through the use of Cultural Support Teams attached to Special Operations units, and Female Engagement Teams engaged in community outreach.[30] Both emphasized local economic and social governance participation and protection of women, which contrasted to their otherwise limited participation in political systems at the time.[31] However, one of the primary challenges of the program was the lack of consistent pre-mission training and harmonization of efforts 'down-range'. So, while US military units included Female Engagement Teams in diverse operations from headquarters' staff to reconstruction and agrobusiness support, the overall effort tended to be ad hoc.[32] Compared with earlier military government operations, Operations Enduring and Iraqi Freedom were plagued by changing mission requirements due to policy shifts from counterterrorism, to counterinsurgency, to nation, and state building. As a

---

29 Two publicly available reviews can be found at https://www.army.mil/article/6496/show_me_the_money_unit_civil_affairs_team_hand_out_10000_in_micro_grants_to_small_business_inst, last accessed 09/05/2022. https://www.sfgate.com/news/article/Disabled-kids-in-a-pinch-get-GI-help-Baghdad-2734183.php.

30 Gayle Lemmon's *Ashley's War: The Untold Story of a Team of Women Soldiers on the Special Operations Battlefield* (New York: Harper, 2015) is widely acknowledged as a definitive review of Cultural Support Teams; for more on the genesis and increasing application of Female Engagement Teams, see *Female Engagement Teams: Making the Case for Institutionalization Based on US Security Objectives in Africa* (US Army Command and General Staff College Press, 2014).

31 Jared Tracy, 'The US Army Cultural Support Team Program,' *Veritas* 21, no. 2 (2016); Matt Pottinger, *et al.* 'Half-hearted: Trying to win Afghanistan without Afghan women,' *Small Wars Journal* (02/8/2010).

32 Janet Holliday, 'Female Engagement Teams: The Need to Standardize Training and Equipment,' *Military Review* (March-April 2012).

result, the potential aggregate impact of Female Engagement Teams was often either muted, lost in the confusion, or rejected outright as inherently hostile to communal values.

Aside from some notable challenges in Iraq and Afghanistan, the overall history of US military government and support to local governance has been relatively consistent, evolving instep with expanded US presence and interests worldwide. This has included an expanded enfranchisement of US women servicemembers, as well as prioritization of women as key stakeholders in governance. The current strategic environment has increased the need for both given the significance of governance as a primary area of competition. In response, and given women's roles as agents, recipients, and wielders of influence, US Civil Affairs has built on the lessons of past operations to expand WPS-aligned capabilities and engagement in support of partners and allies.

Governance and WPS in Geostrategic Competition

US national strategic guidance identifies adversary efforts to undermine confidence in democracy, free markets, open societies, and the current international legal framework. To counter and defeat those threats, institutional coordination and policy implementation relies on common understanding of key concepts. Foremost is the nature of the strategic environment as underpinned by both continuity and change. While identifiable eras, such as the Napoleonic, Industrial, or the Cold War, define the specific character of actors, actions, and arenas, certain elements of competition and conflict remain constant. Whether defined by scarcity of resources, technological innovation to challenge status quo interests, or changes to participation in the system, strategic considerations continue to focus on core priorities. Chief among them are physical security and economic prosperity.

Current US approaches to geostrategic competition seek to harmonize organizational perspectives around two central themes within those goals: populations as essential partners in national security, and governance as a primary arena where influence is won or lost.[33] While such efforts persistently reference 'whole of government' and increasingly 'whole

33 US National Security Strategy (2017), US Interim National Security Strategy (2021), US Joint Strategic Plan: Department of State and Agency for International Development (2022), US National Defense Strategy (2022).

of society' partnerships, the presence of bureaucratic interests (often contradictory or at least self-interested vis-à-vis others) limits the type, breadth, and frequency of internal US government cooperation. The same holds even truer when working with international partners due to the vagaries of policy and contrasting national interests. However, despite those limitations, some areas do show greater common purpose and process.

US strategic guidance emphasizes general approaches to stability and development as ways to increase participation, protect populations, and prevent conflict. The recent US interagency *Stabilization Assistance Review* (2018) defines the importance of those efforts and outlines specific steps to help restore stability in post-conflict areas, and to promote it as a way to prevent future conflict. Concerning US Civil Affairs and the WPS agenda, the guidance prioritizes governance participation as central to stabilization goals. As one of the core tasks of Civil Affairs, and given women's central role in its considerations, the significance of the Review extends beyond doctrine, policy, and resourcing. It also speaks to the underlying US paradigm of military government and the role of women participants, key aspects of previous Civil Affairs operations as much as current ones. The following examples of current US operations support the expansion of women's participation through partnership in governance activities, and efforts to empower communities as part of broader strategic competition against the US's main geopolitical rival – China.

Civil Military Support Element—El Salvador

Decades after El Salvador's civil war (1979-1992), the country struggles with the legacies of violence against civilians and disconnections between communities and the governments that represent them. These problems have contributed to the rise of violent criminal organizations that prey on populations, further delegitimizing governance in the country.[34] As an important regional partner, US involvement in El Salvador has worked to resolve lingering challenges through more trade, counter-narcotics, and organized crime task forces.

Civil Affairs partnerships with local communities and all levels of government have also expanded the role of women in post-conflict

34 Bradford McGuinn, 'Where Murder is Easy: The Fragile State of El Salvador's Liberalism,' in *Fragile States in the Americas*, eds. Jonathan Rosen and Hanna Kassab (New York: Lexington, 2017), 79-102.

development. As an example, in early January 2022, Civil Affairs soldiers coordinated and ran a WPS conference in Tepecoyo, a small city in the western region of the country.[35] The four-person Civil Affairs Team worked with local women's groups to identify areas of concern and ways to increase participation in policy making. The team facilitated classroom instruction on entrepreneurship, leadership development, and health education. The event emphasized the promotion of women's involvement in advancing economic opportunities for local communities as part of broader post-conflict stabilization.

The Tepecoyo conference also contributed to US diplomatic efforts by supporting the furtherance of El Salvador's WPS National Action Plan. Subsequent engagements occurred through a joint US Civil Affairs – Agency for International Development seminar on women in security. Local officials and women's groups participated with El Salvadoran National Police and Armed Forces leadership to address policing and military roles at the community level. The interagency program supported building partner capacity through increased recruitment and the training of women, methods to message needs and successes through community relations, and support to local governance implementing the same.

The Civil Affairs operations did not directly target China's activities in the country. Yet through their support of local governance and national WPS initiatives, these multi-level engagements helped to align policies and signal importance and US commitment to El Salvador. Taken together, US engagement with El Salvadoran governance contributes to regional US efforts countering China's predatory influence by expanding participation and strengthening democratic legitimacy.[36]

Combined Joint Task Force—Horn of Africa

The only permanent US base in Africa, Camp Lemonnier, is in Djibouti. The area has become a contested operational space as China's naval presence increases under the guise of commercial port investment. As so often occurs with the Belt and Road Initiative though, local market development does

35 See the following publicly available reference, 'US Special Operation Forces Civil Affairs Enables WPS Event in El Salvador,' https://dialogo-americas.com/articles/us-special-operation-forces-civil-affairs-enables-wps-event-in-el-salvador/.

36 See Evan Ellis, 'China and El Salvador: An Update,' *CSIS* (March 2021) for the rise of China's economic and political influence, and implications for neighboring partnerships with the US.

not follow initial promises.[37] The anticipated local economic booms have concentrated benefits for China's state-owned enterprises and military development. More damaging for Djibouti has been the predatory lending practices that include surrender of port ownership on condition of default, often at very high interest rates.[38] While Chinese debt policies towards African debtor nations have eased somewhat, the trajectory remains extractive rather than cooperative.[39]

US Civil Affairs soldiers have facilitated English language courses to broaden the opportunities available to local communities. Courses began in May 2021 and not only covered how to speak English, but also highlighted different international avenues available to communities seeking foreign direct investment. The courses also covered business applications in coordination with local jobs training programs, specifically for women participating in commercial markets. Civil Affairs classroom instruction coincided with veterinary exchanges with local farming and herding communities, and medical information sharing at one of the main hospitals in Djibouti, highlighting the key role of women in agriculture and family care.[40]

Such efforts to expand opportunities to develop and utilize market connections, as well as educational opportunities to improve basic healthcare directly support WPS goals. In particular, the working groups and classroom instruction emphasized identifying the types of social capital women bring in areas of food security and health, and ways to message relevant government decision makers with local needs. Thus, despite the small investment compared to Chinese infrastructure development in the city, Civil Affairs engagement has helped to legitimize a broader US

37 'Data: Chinese Workers in Africa 2009-2020,' *China-Africa Research Initiative—Johns Hopkins University School of Advanced International Studies*, https://www.sais-cari.org/data-chinese-workers-in-africa. Jennifer Hillman and David Sacks, 'China's Belt and Road: Implications for the United States,' *Council on Foreign Relations* (March 2021), https://www.cfr.org/report/chinas-belt-and-road-implications-for-the-united-states/#main-content.

38 Venkateswarn Lokanathan, 'China's Belt and Road Initiative: Implications for Africa,' *Observer Research Foundation* (August 2020) offers an effective review of BRI, despite its many supporters and detractors.

39 Yun Sun, 'China and Africa's debt: Yes to relief, no to blanket forgiveness,' *Brookings* (April 2020), https://www.brookings.edu/blog/africa-in-focus/2020/04/20/china-and-africas-debt-yes-to-relief-no-to-blanket-forgiveness/.

40 See for public release, 'CJTF-HOA hosts inaugural Women in Security Symposium in Djibouti,' https://www.africom.mil/article/34059/cjtf-hoa-hosts-inaugural-women-in-security-symposium-in-djibouti and 'US Army Civil Affairs works to empower women in Djibouti,' https://www.africom.mil/image/33858/us-army-civil-affairs-works-to-empower-women-in-djibouti.

message of partnership, rather than extraction from local populations, with women as key social actors.[41]

INDOPACOM—Mongolia Women's Mentorship Program

Mongolia has some of the world's worst air quality. Its capital city, Ulaanbaatar, consistently ranks among the dirtiest due to coal burning for domestic heating and Chinese-developed commercial power generation.[42] Outside of the capital, the country's rapidly expanding mining operations often use open-air tailing ponds to hold poisonous chemicals that leach into the soil, contaminating ground water, to say nothing of the toxic fumes across the surrounding area.[43] Combined with increasing urbanization of the capital city, the loss of fresh water supplies and agricultural farmlands stress governance capabilities. Communities caught in this development/pollution trap struggle to have a voice in the macroeconomic forces shaping their lives.

USINDOPACOM, the four-star military command focused on Asia and the Pacific regions, has used joint training exercises with the Mongolian Defense Ministry to address some of those issues. One in particular, the Mongolian Women's Mentorship Program held in September 2019, collaborated with US diplomatic and development partners on ways to increase women's participation in Mongolian environmental and disaster relief efforts.[44] The event included leadership training for rural women and advocacy groups seeking to redress governance problems, in part resulting from pollution in the capital and countryside. Coinciding with Mongolian government efforts, the US military engagements have increased women's

---

41 USAFRICOM maintains a broad WPS effort, including continued work with local, regional and federal government officials and community leaders to develop Djibouti's National Action Plan (April-May 2022), as well as through partnership on Health Security Synchronization with the Moroccan Armed Forces (June 2022), and participation in the African Military Law Forum in Botswana (August 2022).

42 D. Amarsaikhan, *et al.* 'A Study on Air Pollution in Ulaanbaatar City, Mongolia,' *Journal of Geoscience and Environment Protection*, 2, (2014): 123-128. Jargalsaikhan Bardach, *et al.* 'Winter Air Pollution from Domestic Coal Fired Heating in Ulaanbaatar, Mongolia, Is Strongly Associated with a Major Seasonal Cyclic Decrease in Successful Fecundity,' *International Journal of Environmental Research and Public Health*, 18, no.5 (March 2021): 2750-2761.

43 Clement Dai and David Gibson, 'Minegolia: China and Mongolia's Mining Boom,' *Wilson Center China Environment Forum* (July 2013): 1-11. Adrian Gheorghe, *et al.* 'Mongolia's Air Pollution Crisis: A Call to Action to Protect Children's Health,' *UNICEF* (February 2018).

44 Sharon Feist, 'Women Key to Peace and Security in Mongolia and Globally,' illustrates many of the principles and programs used within the INDOPACOM area of operations, https://ipdefenseforum.com/2019/10/women-key-to-peace-and-security-in-mongolia-and-globally/.

participation through senior-level discussions directly effecting local communities and national priorities.[45] While not exclusively Civil Affairs-led, the USINDOPACOM event shows a high level of senior military leader commitment towards the WPS agenda, and its benefits to governance in an area of geostrategic importance.

The above examples demonstrate ways US Civil Affairs embeds the pillars of the WPS, especially participation, as part of US geopolitical competition with China, but they did not come about spontaneously. They developed over time and represent a deliberate effort to meet US strategic objectives. Including a plurality of voices and venues for participation remains a central theme, as does the role of US military support for it. Civil Affairs helps to fill that role, so long as the force is equipped, trained, and educated to navigate the complexities of the civil environment, including the necessity of women as key participants in it.

Educating Future Civil Affairs Forces: Governance Studies and WPS

As the US national security strategy has shifted focus from counterterrorism to emerging threats from great power adversaries, US Civil Affairs forces have stepped back from the high operational tempo of the Global War on Terror. With more 'dwell time' away from the battlefield, soldiers and staff have gained greater opportunities to evaluate nearly two decades of reporting on operational effectiveness. The renewed focus on modernizing and aligning capabilities for geostrategic competition reveals a broad gap in foundational governance knowledge, despite nearly two decades of extensive governance support activities abroad. As a result, the US Army Special Warfare Center and School revised its governance curriculum in 2019, focusing on governance as a battlefield for influence and identifying participation as a primary mechanism to increase the legitimacy of governance.[46]

With the addition of more than 200 hours focused on local networks, formal and informal rules, normative foundations, and organizational structures, the new curriculum relies on an interdisciplinary approach to

45 'Women's Resilience in Mongolia: How Laws and Policies Promote Gender Equality in Climate Change and Disaster Relief Management,' *Asian Development Bank* (June 2022): 1-52.
46 Personal involvement as governance advisor and course curriculum developer.

the civil environment and its role in geostrategic competition. It aims to prepare Civil Affairs soldiers for the challenges of knowing whom to engage, how to do so, and most importantly, why it matters. Building effective partner governance promotes democratic participation and helps to protect civilian populations, both of which are in the US national interest. Using multidisciplinary course modules developed from comparative politics, sociology, narrative discourse analysis, and economic data, civilian and military instructors analyze previous Civil Affairs operations to identify operational decision points that either successfully engaged key stakeholders, or missed opportunities and the reasons for it. The combined five-weeks of instruction fosters better awareness of the civil environment 'facts on the ground' for later mission planning and operational execution. Equally importantly, it teaches students to understand, analyze, and communicate 'lessons learned' to their commanders and local communities, broadening the impact of good governance approaches and increasing operational effectiveness over time.

Women, Peace and Security goals are woven throughout the curriculum. The approach emphasizes the connection between participatory governance and WPS by revealing the role of women across operational environments, rather than limit it to specific topics that 'check the block'. In conjunction with long-term increases in women soldiers across the US military, the course 'widens the aperture' of governance support from technical means for physical infrastructure, to include diverse perspectives about legitimacy and how it impacts different groups. Learning to listen to different voices, in particular women as governance participants, benefits WPS goals. It also enables Civil Affairs soldiers to partner with key stakeholders with greater understanding of local contexts through increasing gender awareness and coherence.[47] This will enable the future force to assist partner governance more effectively as it enhances women's participation writ large. In that regard, the new Civil Affairs course advances the longstanding priorities of US military government and Women, Peace and Security goals at home and abroad

47 US Department of Defense, WPS Strategic Framework and Implementation Plan (2020) 12-14.

References

Althalathini, D., *et al.* 'The Impact of Islamic Feminism in Empowering Women's Entrepreneurship in Conflict Zones: Evidence from Afghanistan, Iraq, and Palestine,' *Journal of Business Ethics,* (2022) 39-55.

Amarsaikhan, D. *et al.* (2014). A Study on Air Pollution in Ulaanbaatar City, Mongolia. *Journal of Geoscience and Environment Protection,* (2014) 123-128.

Asian Development Bank, 'Women's Resilience in Mongolia: How Laws and Policies Promote Gender Equality in Climate Change and Disaster Relief Management,' (June 2022) 1-52.

Bandiera, L., *et al. Jobs in Iraq: A primer on Job Creation in the Short-Term.* World Bank, 2019.

Bardach, J., *et al.* 'Winter Air Pollution from Domestic Coal Fired Heating in Ulaanbaatar, Mongolia, Is Strongly Associated with a Major Seasonal Cyclic Decrease in Successful Fecundity,' *International Journal of Environmental Research and Public Health,* (March 2021) 2750-2761.

Carruthers, S., *The Good Occupation: American Soldiers and the Hazards of Peace.* Cambridge: Harvard University Press, 2016.

China-Africa Research Initiative, 'Data: Chinese Workers in Africa 2009-2020,' *Johns Hopkins University School of Advanced International Studies,* (2021).

Dai, C. and David Gibson, 'Minegolia: China and Mongolia's Mining Boom,' *Wilson Center China Environment Forum,* (July 2013).

Ellis, E., 'China and El Salvador: An Update,' *CSIS* (March 2021).

Gheorghe, A., *et al.* 'Mongolia's Air Pollution Crisis: A Call to Action to Protect Children's Health,' *UNICEF* (February 2018).

Henning, E. and Rangwala, G., *Iraq in Fragments.* London: Hurst, 2006.

Hillman, J., and David Sacks, 'China's Belt and Road: Implications for the United States,' *Council on Foreign Relations,* (March 2021).

Holliday, J., 'Female Engagement Teams: The Need to Standardize Training and Equipment,' *Military Review* (March-April 2012) 90-94.

Hudson, V., Bowmen D., and Perpetua Neilsen, 'Clan Governance and State Stability: The Relationship between Female Subordination and Political Order,' *American Political Science Review,* (2015) 535-55.

Graves, W., *America's Siberian Adventure: 1918-1920.* American History Classics, 2019.

Lemmon, G., *Ashley's War: The Untold Story of a Team of Women Soldiers on the Special Operations Battlefield.* New York: Harper, 2015.

Lokanathan, V., 'China's Belt and Road Initiative: Implications for Africa,' *Observer Research Foundation* (August 2020) 1-11.

McGuinn, B. 'Where Murder is Easy: The Fragile State of El Salvador's Liberalism.' In *Fragile States in the Americas,* edited by Jonathan Rosen and Hanna Kassab, 79-102. New York: Lexington, 2017.

Maulucci, T. and Junker, D., *GIs in Germany: The Social, Economic, Cultural, and Political History of the American Military Presence.* Washington: Cambridge University Press, 2013.

Nunnally, M., *American Indian Wars: a Chronology of Confrontations between Native Peoples and Settlers and the United States Military*. London: MacFarland and Co., 2007.

Pottinger, M. *et al*. 'Half-hearted: Trying to win Afghanistan without Afghan women,' *Small Wars Journal*, (February 8, 2010).

Rathmell, A., 'Planning Post-Conflict Reconstruction in Iraq: What Can We Learn,' *International Affairs*, (2005) 1013-38.

Sandler, S., *Glad to See Them Come and Sorry to See them Go: A History of US Army Tactical Civil Affairs/Military Government, 1775-1991*, 1994.

Shadlow, N., *War and the Art of Governance: Consolidating Combat Success to Political Victory*. Washington: Georgetown University Press, 2017.

Stirk, P., *The Politics of Military Occupation*. Edinburgh: Edinburgh University Press, 2009.

Statement of LTG Johnathan Braga, Commanding General, USASOC Before the Senate Armed Services Committee, (April 2022).

Tracy, J., 'The US Army Cultural Support Team Program,' *Veritas*, 2016.

Treadwell, M., *The Women's Army Corps*. Washington: US Army Publication, 1991.

US Army, *American Military Government of Occupied Germany: 1918-1920*. 1943.

US Army, *Female Engagement Teams: Making the Case for Institutionalization Based on US Security Objectives in Africa*. US Army Command and General Staff College Press, 2014.

US Army Center of Military History, 'The Mexican War and After,' *American Military History*, 1989.

US Army Field Manual 3-0, *Operations*, 2017.

US Army Field Manual 3-07, *Stability*, 2014.

US Army Field Manual 3-24, *Insurgencies and Countering Insurgencies*, 2014.

US Army Field Manual 3-57, *Civil Affairs Operations*, 2021.

US Army Field Manual 41-5, *Joint Manual for Civil Affairs*, 1966.

US Army, *People Strategy – Diversity, Equity and Inclusion Annex*, 2020.

US Army TRADOC *Gender Integration Study*, 2015.

US Department of Defense Directive 2000.13, *Civil Affairs Operations*, 2017.

US Department of Defense Joint Doctrine Note 1-19, *Competition Continuum*, 2019.

US Department of Defense Joint Publication 1-02, *Joint Operating Concept—Irregular Warfare: Countering Irregular Threats*, 2010.

US Department of Defense Joint Publication 1-02, *Department of Defense Dictionary of Military and Associated Terms*, 2008.

US Department of Defense Joint Publication 3-13, *Information Operations*, 2014.

US Department of Defense, *Summary of the Irregular Warfare Annex to the National Security Strategy*, 2020.

US Department of Defense, *Women, Peace, and Security Strategic Framework and Implementation Plan*, 2020.

US Department of Defense, *Women, Peace, and Security Framework and Implementation Plan*, 2020.

US Department of State, *Stabilization Assistance Review*, 2018.

US Government, *Strategy on Women, Peace, and Security*, 2019.

US Government, *Women, Peace, and Security Congressional Report*, 2022.
US Interim National Security Strategy, 2021.
US Joint Strategic Plan: Department of State and Agency for International Development, 2022.
US National Security Strategy, 2017
US National Defense Strategy, 2022.
US Public Law 115-68, *Women, Peace, and Security Act*, 2017.

3

UN PEACEKEEPING AND THE BRITISH ARMY

Steve Maguire

In 1994 a UN Special Representative, Yasushi Akashi, commented that 'boys will be boys' when responding to accusations of sexual abuse during a peacekeeping operation in Cambodia.[1] Such perspectives remain common across international society despite decades of discussion over gender equality and conflict related sexual violence (CRSV). For peacekeeping missions, this was especially true prior to the adoption of the UN Security Council Resolution (UNSCR) 1325 in October 2000. The Resolution 'recognised' the need to mainstream gender perspectives in peacekeeping operations and 'reaffirmed' the importance of women in peacekeeping and post-conflict resolution. UNSCR 1325 formalised the Women, Peace and Security (WPS) agenda which had been developing since the 1990s. The agenda is defined by both subsequent UNSCRs and the literature which has sought to reform peacekeeping and achieve the intent of UNSCR 1325. This chapter looks at UN peace operations as defined in their 2008 capstone doctrine to examine the impact of the WPS agenda in contemporary peacekeeping operations.[2] It uses the term 'peacekeeping' as a catchall phrase to cover the spectrum of UN peace operations. This reflects the 'multi-dimensional nature' of contemporary UN operations and covers the boundaries between conflict, conflict prevention, peace-making, peacekeeping, peacebuilding, and peace enforcement, especially as the UN doctrine views these distinct roles as being increasingly merged.

1 "Profile: Yasushi Akashi", *Independent*, 30 April 1994.

2 UN, UN Peacekeeping Operations Principles and Guidelines (2008), https://www.un.org/ruleoflaw/files/Capstone_Doctrine_ENG.pdf

This chapter demonstrates the impact that the WPS agenda has had on UN operations and relates the conclusions to modern British Army experience.

Masculine perspectives, which view women as victims of war, are common across international society and have been evident in peacekeeping and other military operations. This chapter presents a core argument that the WPS agenda impacts military peacekeeping operations by challenging masculine security perceptions and forcing conceptual and doctrinal evolution. The argument builds on Elizabeth Defeis's 2008 conclusion that there was a 'decade of inaction' immediately after 1325 was adopted.[3] Refreshing her conclusion, it will be argued that whilst there is evidence of an acceleration of change since 2009, the impact of which has not yet been institutionalised. Looking internationally, the WPS agenda has been received differently and applied with mixed depth dependent on national gender perspectives. The WPS agenda has not always been embraced by male writers or practitioners fully, suggesting that its impact is concentrated in feminist academia and specialised policy circles.

To demonstrate the agenda's impact on peacekeeping, this chapter adopts a framework presented by Judith Stiehm. Writing in 2001, Stiehm identified 'three 'I's' with which the WPS agenda needed to engage to change peacekeeping missions: inertia, implementation, and institutionalisation.[4] She argued firstly that the WPS agenda has challenged masculine perspectives of peacekeeping highlighting the need for change. Yet a level of strategic inertia remains with institutions and nations focusing on popular WPS topics over more structural and measurable impacts. Secondly, that the international community has different political appetites for the implementation of the WPS agenda based on national gender perspectives. Thirdly, that the UN has worked hard to institutionalise the WPS agenda, particularly through the Department of Peace Operations, but that it needs more time before it is fully embedded. This suggests that masculine responses are often prioritised over engaging with WPS ideas that could lead to success in peacekeeping missions. These factors combine to show a shallow and fragile understanding of the WPS agenda which is often separated from the reality of peacekeeping missions.

---

3 Elizabeth Defeis, "U.N. Peacekeepers and Sexual Abuse," *Washington University Global Studies Law Review* 7, 2 (2008), https://openscholarship.wustl.edu/cgi/viewcontent.cgi?article=1109&context=law_globalstudies.

4 Judith Stiehm, "Women, Peacekeeping and Peacemaking" in *Women and International Peacekeeping*, Louise Olsson and Torunn Tryggestad ed (London, Frank Cass, 2001), 44.

The core of this chapter provides an academic analysis of the current state of WPS in peacekeeping. This chapter, however, also has a second strand which looks to explore what these conclusions mean for the British Army. There is a renewed interest in peacekeeping within the UK's strategic and defence policy. The 2015 National Defence and Security Review declared that "the UK will continue to champion reform to increase the efficiency and impact of UN engagement. We will double the number of military personnel that we contribute to UN peacekeeping operations."[5] Since 2003 the British Army has been actively engaged in operations from combat in Iraq and Afghanistan to peacekeeping in Mali, a continued presence with NATO in the Balkans, and deterring Russia with forward presence in Estonia and Poland. To assess if the general trends identified apply to the British Army, this chapter draws out key implications and relates these points to the Army's military operations. The analysis on UN peacekeeping is general, and how the British Army have applied and can apply these is specific. In many ways, the British Army is ahead of the trends seen in the academic literature. There is evidence that the WPS agenda is having a positive impact on British Army operations, both with conceptual and doctrinal changes as well as reports from missions showing the force enacting WPS principles. There are, however, things which can be done to exploit this momentum and this chapter concludes with eight recommendations to make the UK's WPS commitments a reality in military operations.

Inertia

The adoption of the WPS agenda mandated institutions to develop processes to mainstream gender perspectives in peacekeeping missions. However, the impact is limited, and changes often appear more cosmetic than substantial, suggesting that a degree of inertia remains. The UN first defined gender mainstreaming in 1997 as the process of "assessing the implications for women and men of any planned activity" showing that the importance of gendered awareness in peacekeeping was embedding before UNSCR 1325.[6] However, Gina Heathcote notes that the UN Security Council still makes gendered assumptions undermining the agenda.

5 HMG, National Security Strategy and Strategic and Security Review 2015, https://www.gov.uk/government/publications/national-security-strategy-and-strategic-defence-and-security-review-2015.

6 UN, Report of the Economic and Social Council for the Year 1997, A/52/3/Rev.1.

Heathcote points to the "adoption of strategies for counting women", rather than mechanisms for transforming gender relations, as evidence of this inertia.[7] UN Women reports into the implementation of WPS have a greater focus on the number of women engaged and often struggle to evidence more substantial points beyond generalist language.[8] Heathcote argues that the UNSC has not identified men as a gendered actor which means that the WPS agenda fails to fully challenge deeply embedded assumptions concerning gender roles.[9] However, this critique must also be balanced against time. The mandates of the missions deployed post 2000 all 'reaffirm' the principles of 1325, demonstrating an impact on mission design.[10] This suggests that the conceptual basis of peacekeeping has been challenged by the WPS agenda, but more time is needed to develop and clarify what practical changes are needed to further challenge inertia in the system.

One reason explaining why depth of impact has not been fully realised is related to ideas of war and peace stemming from deeply embedded masculine perspectives about women as passive victims of and men as active participants in conflict. Since 2000, the WPS agenda has challenged these perceptions and impacted peacekeeping missions by developing supporting concepts. Linking to Heathcote's argument, Gabrielle Simm argued in 2009 that there was a 'fear' in the UNSC that associating with the WPS agenda would "weaken its symbolic power as a 'masculine institution'".[11] The underlying assumption of this position is that only masculine bodies are capable of restoring peace which demonstrates a narrow and masculine view of peace as the absence of war. This highlights a high level of inertia in understanding. Chantal de Jonge Oudratt pointed out in 2013, the "lack of strong systematic empirical evidence" means it is difficult to challenge these perceptions as it relies on soft concepts that are not regularly accepted by traditional masculine institutions such as the UNSC.[12] Such viewpoints do not fully acknowledge the context of how

---

7 Gina Heathcote, "Participation, Gender, and Security", in *Rethinking Peacekeeping, Gender Equality and Collective Security*, Gina Heathcote & Dianne Otto ed (Basingstoke, Palgrave Macmillan, 2014) 50. See also chapter 1 of this collection.

8 See for example UN Women, "Preventing Conflict, Transforming Justice, Securing Peace", http://peacewomen.org/sites/default/files/UNW-GLOBAL-STUDY-1325-2015%20(1).pdf.

9 Heathcote, "Participation, Gender and Security" 50.

10 UN, "Peacekeeping", https://peacekeeping.un.org/en/mission. 12 missions were active in September 2022.

11 Gabrielle Simm, *Sex in Peace Operations* (Cambridge, Cambridge University Press, 2013) 37.

12 Chantal de Jonge Oudraat, "UNSCR 1325 – Conundrums and Opportunities", *International Interactions* 39, 4 (2013) 612-619.

the debate has developed since 2000. Looking for data-driven answers can miss the more nuanced impacts of the WPS agenda and essentialises women and men into stereotyped categories, and that is before considering other aspects of gender. For example, a 2015 UN report noted that previously masculine definitions of negative peace, which guide missions, as the 'absence of war' have been expanded and women are now considered 'central stakeholders in bringing about positive peace'.[13] British Army gender expert Colonel Rosie Stone notes that attitudes towards women in security have changed because of the increasing participation and visibility of women.[14] Such changes would not be possible if the debate had remained fixed in masculine perceptions. These factors combine to show that the WPS agenda is actively challenging inertia within the peacekeeping system and is impacting the conceptual basis of peacekeeping missions.

One of the effects on peacekeeping operations has been a focus on CRSV at the expense of the broader aims of the WPS agenda. The inability to discuss WPS holistically shows a level of inertia within the debate. This has reinforced perceptions of women as victims and not proactive security actors thereby undermining the effectiveness of peacekeeping missions. It has also created a political mechanism for states to avoid broader challenges to masculine positions. Karen Eagle identifies that there is a 'relentless attention' to sexual violence at the expense of other agenda issues.[15] She points to the use of 'celebrity diplomacy' as the key reason that sexual violence has narrowed the WPS agenda since 2014. Eagle uses the example of the 'Vagina Monologues', a campaign created by Eve Ensler, as having an international impact by utilising celebrities such as Nicole Kidman to gain support. Drawing on Simm's deductions previously mentioned, this plays to masculine stereotypes of women in conflict needing protection and downgrades them to victims of conflict framing the debate around peacekeeping.[16]

Demonstrating how this can narrow the debate, the Global Network of Women Peacebuilders noted in an open letter that UNSCRs on peacekeeping are still passed 'without consultation' with women relying

---

13 UN Women, "Preventing Conflict", 25.

14 Rosie Stone, "The Third Sex", *Wavell Room* (2019), https://wavellroom.com/2019/01/29/the-3rd-sex-gender-and-contemporary-conflict/.

15 Karen Eagle, "The Grip of Sexual Violence", in *Rethinking Peacekeeping*, 23.

16 Simm, *Sex in Peace Operations*, 37.

instead on traditional political structures.[17] Going further, they note that only recognising women as 'victims as war' denies agency and they call for greater recognition of women as participants in line with the broader WPS agenda. The contrast of this position of empowerment with the more popular victim focused narrative is stark. It may be that the UN needs more understanding of the societies into which they deploy missions, to identify patriarchal cultural traits where law and order has broken down, or how violence manifests against men, or maybe more political courage is needed to confront these attitudes. It does suggest that there needs to be a deeper consideration of how the UN wishes to enable the peaceful resolution of conflict and how to utilise peacekeeping missions to achieve these aims. This is important because these debates impact the considerations of the UNSC and shape the mandates of peacekeeping missions. The WPS agenda can challenge any political inertia. As Judith Steihn noted, mission mandates reflect the debate and understanding of the practical application of the WPS agenda.[18] The mandate guides the mission commander and reminds them of their responsibilities and gives legitimacy, or not, to objectives. Narrowing the debate to focus on CRSV enables masculine perspectives to maintain the control of the narrative that women are victims of war without engaging with the full WPS agenda undermining mission success. A narrow debate does not reflect the entirety of the WPS agenda but does show a level of inherent inertia in discussing what to do and how to enact change.

Inertia and the British Army

To some extent, the British Army mirrors this high level of inertia evident with UN peacekeeping but is showing an acceleration of change and is overcoming many problems. In 1999, the then Minister for the Armed Forces declared "every day is International Women's Day in the modern Armed Forces".[19] That it took another 17 years to remove the barrier to women serving in combat roles, despite growing evidence of effectiveness in combat, and to source fitting protective equipment, suggests there is a

17 Global Network of Women Peacebuilders, "Open Letter to Member States of the Security Council Re: Res 1960", https://www.peacewomen.org/content/letter-open-letter-member-states-security-council-re-res-1960 (2011).
18 Stiehm, "Women, Peacekeeping and Peacemaking", 45.
19 Rachel Woodward and Patricia Winter, "Discourses of Gender in the Contemporary British Army." *Armed Forces & Society* 30, 2 (2004): 279–301.

level of embedded inertia in the British Army. Anecdotal commentary from within the British Army bemoans an attitude that recognises the importance of women but concludes that the Army simply does not care enough.[20] The 2021 Atherton Report into women in the Armed Forces suggests that very little has changed.[21] Linking to Chantal de Jonge Oudratt's argument about a need for hard evidence, Joanne Mackowski noted that women were allowed to enter combat roles despite a lack of such evidence.[22] Arguably, societal and generational changes have proved more effective at challenging inertia than the Army's ability to challenge its own embedded social expectations.

Wider social changes alone, however, do not mean that the Army is immune from inertia. Much of the commentary mentioned above is about the internal culture of the British Army and not how it operates when deployed. But the two issues are explicitly linked. The public debate is equally fixed on gender being separate from how a force operates and this has unfairly given the impression that the British Army has a high level of inertia. In 2007, fifteen Royal Navy personnel were captured by the Iranians. The capture of a woman, Leading Seaman Faye Turney, meant the coverage of her capture became gendered with a focus on her overwhelming 'wish of a mother to be with her child'. This was the public perception of a woman deployed on operations and the death of women soldiers receives similar gendered coverage. An Army officer, Alan Hynes, counters this narrative by arguing that the alternative is a complete removal of women from the Armed Forces denying their right to fight.[23] Hynes "reminds the reader to be aware of them, for they are tools of the regressively insane. This is just but one example". It is clear then that there is a significant difference between the public and academic perceptions of inertia in the British Army and how it might apply the WPS agenda.

A desire for hard evidence to support change remains. The Operation HERRICK Campaign Study noted that "more research is required

---

20 James Burton, "Cultural Apathy and Dishonesty within the British Army", *Wavell Room* (2020) https://wavellroom.com/2020/12/09/cultural-apathy-and-dishonesty-within-the-british-army-say-do-gap/.

21 House of Commons Defence Committee, *Protecting Those Who Protect Us: Women in the Armed Forces from Recruitment to Civilian Life*, 25 July 2021, https://committees.parliament.uk/publications/6959/documents/72771/default/.

22 Joanne Mackowski, "Combat Inclusion: The End of Gender Segregation in the British Armed Forces", July 11, 2016, https://rusi.org/explore-our-research/publications/commentary/combat-inclusion-end-gender-segregation-british-armed-forces.

23 Alan Hynes, "Progressive Insanities: Women in Defence", *Wavell Room* (2021), https://wavellroom.com/2021/04/20/progressive-insanities-women-in-defence.

to determine whether women soldiers should be force generated in engagement-specific roles at the cost of other capabilities".[24] This need for data to inform decision making clearly did impact how the British Army failed to fully consider the WPS agenda. The HERRICK Campaign Study also does not mention the WPS agenda despite its maturity when the study was published in 2015. Yet, despite this, there is now evidence of accelerating change in the British Army's approach. Writing in February 2022, the Commanding Officer of the UK contingent to the UN Mission in Mali noted the 'vital' importance of the WPS agenda in being effective.[25] The wider text demonstrates a far broader appreciation of WPS objectives than much of the commentary published since 2003. Story boards from the same mission also detail how the UK contingent enabled UN training events.[26] Released media coverage focuses on the role of women as professional soldiers reaching out to local communities challenging embedded public perceptions of women as victims of war. In another example, the British Army was key in teaching skills needed to enact the WPS agenda through training courses delivered to the Ukrainian Army covering all ranks and multiple combat specialists.[27]

This demonstrates that the British Army is actively engaged in delivering the WPS agenda and addressing some of the specific challenges faced by women and children in conflict. The contrast with a decade of campaigning in Afghanistan and the conduct of operations since 2015 shows this change. The British Army's mentality and core ideology has changed in line with the wider debate and is now being deployed on missions. These examples do not change the more strategic themes prominent in the academic literature, but they do show that there is a growing emphasis in the British Army on challenging inertia and implementing the WPS agenda.

---

24 Ministry of Defence, Op HERRICK Campaign Study (2015), 5-7_3, https://assets.publishing.service.gov.uk/government/uploads/system/uploads/attachment_data/file/492757/20160107115638.pdf.

25 Hugo Lloyd, "British Army in Mali", *Wavell Room* (2022), https://wavellroom.com/2022/02/11/britisharmy-in-mali/.

26 British Army story boards from Mali 2022. UN Confidential document held by author.

27 Ministry of Defence, Joint Service Publication 1325 Part 2: Guidance (2019) https://assets.publishing.service.gov.uk/government/uploads/system/uploads/attachment_data/file/770922/JSP1325_Part2.pdf.

Implementation

The WPS agenda has increased the number of women now involved in peacekeeping missions. In 2015 Louise Olsson et al, collated UN data to show that the number of women in peacekeeping had risen to 3,753, or 4 per cent of peacekeeping forces.[28] Updating these figures in 2020, the UN point to 4.8 per cent of military and 10.9 per cent of police contingents being women demonstrating a slow but continuous rise.[29] Whilst this remains a minority of peacekeepers, it represents a 22 per cent increase in women deployed using 2009 as a baseline. Looking to exploit this momentum, the UN has set ambitious targets for increasing women's participation in missions to 15 per cent of military and 20 per cent of police contingents. In part, however, this increase is explained because of the 2008 peacekeeping doctrine which broadened the definition of peacekeeping missions to include police elements, widening the pool of women available to be counted. However, despite this impact, the overarching detail hides a general trend that missions which are considered dangerous have low women's participation and most of this increase is in 'safe' missions.[30] Kajsa Tidblad-Lundholm develops this argument by suggesting that troop contributing nations are less likely to deploy women during the early stages of deployment or to missions with high levels of operational risk.[31] Her research also notes that women's participation in missions increases the longer the mission is deployed. One potential counter argument is that the identified rise reflects the growing number of women choosing to join national militaries and the removal of barriers to serving in combat arms. The wider research of Olsson et al, however, shows that there is no clear link between increased representation and the number of women deployed on peacekeeping missions in NATO countries, and this trend is likely to be global.[32] This suggests that masculine security perspectives

---

28 Louise Olsson, Anita Schjølse, Frida Möller, "Women's participation in international operations and missions", in *Gender, Peace and Security*, Theodora-Ismene Gizelis & Lousie Olsson ed (Abingdon, Routledge, 2015), 44.
29 UN, "Women in Peacekeeping", https://peacekeeping.un.org/en/women-peacekeeping.
30 Louise Olsson, Anita Schjølse, Frida Möller, "Women's participation in international operations and missions", in *Gender, Peace and Security*, Theodora-Ismene Gizelis & Lousie Olsson ed (Abingdon, Routledge, 2015).
31 Kajsa Tidblad-Lundholm, "When Are Women Deployed?", *International Peacekeeping* 27, 4 (2020), 673-702.
32 Olsson, Schjølse, Möller, "Women's Participation in Missions", in *Gender, Peace and Security*, 46-48.

remain embedded and that implementation of the WPS agenda is fixed in national gender perspectives.

One practical impact of the WPS agenda has been the adoption of mandatory gender training for peacekeepers, but this training has not been sufficient to meet the UN objectives. UNSCR 1325 called specifically for "specialised training…[on] the special needs and human rights of women and children". Progress in developing the needed training packages, however, has been slow. Questioning the quality of gender training in 2006, Karen Barnes noted that training was largely "tokenistic gestures that contradict the essence of what mainstreaming seeks to achieve".[33] A UN report in 2010 found that more training was needed to 'integrate gender perspectives' suggesting that training from 2000 to 2010 was not sufficient to achieve basic impacts.[34] By 2019, gender-related training only accounted for 2 hours in 10 days of pre-deployment training for peacekeeping demonstrating its relative importance to other training objectives.[35] Critiquing the UN for this failure, to some extent, removes a burden of responsibility from troop contributing states and suggests they do not have the same level of responsibility for enacting the WPS agenda. In part, this is likely to be because of low levels of women in representative positions across the international security sector and more general inertia in implementing the WPS agenda. However, the UN's training package has also been considered 'too rigid' to be effective.[36] People, including peacekeepers, learn best by doing, not watching. It may also be the case that a higher level of courage is needed to engage with potentially sensitive gender-related issues. The need to maintain consensus could be behind training which is not considered sufficient to give peacekeepers the information they need for military effectiveness. There is also a need for collective training alongside individual education courses. As such, it is likely that the UN needs to change its approach to training to implement the WPS agenda if it wants to fully embed it into modern operations.

Broader social issues also impact this debate. Sabrina Karim and Kyle Beardsley identify a gendered protective norm within military communities

---

33 Karen Barnes, "Reform of More of the Same?", YCISS Working Paper 41 (2006), https://genderandsecurity.org/sites/default/files/Barnes_-_Reform_or_more_of_the_same_0.pdf.
34 UN, "Ten Year Impact Study on UNSCR 1325", https://www.globalpolicy.org/images/pdfs/10year_impact_study_1325.pdf.
35 Marina Caparini, "Gender Training for Police Peacekeepers", https://www.sipri.org/commentary/topical-backgrounder/2019/gender-training-police-peacekeepers-approaching-two-decades-united-nations-security-council
36 IPI roundtable, New York, 24 January 2020.

which degrades attempts to train the WPS objectives.[37] The impact of the WPS agenda on peacekeeping training has been shaped by masculine norms undermining its potential effectiveness. It has failed to adequately challenge embedded gender biases inherent in organisations dominated by men or masculine culture. For example, Georgina Holmes presents evidence that suggests Rwanda did not adequately prepare women peacekeepers because of a baseline assumption that women already possessed adequate soft skills.[38] Karim and Beardsley's wider research has found that many potential women peacekeepers were put off volunteering because of the pressures of family life, social expectations, and a lack of gender specific support.[39] This deduction flows from the high level of strategic inertia and embedded masculine norms identified above. It suggests that the WPS agenda is impacting peacekeeping conceptually but is struggling to achieve practical impacts on missions.

The most common way for states to implement the WPS agenda is through the adoption of national action plans (NAPs) but the impact has been mixed and is defined by national gender perceptions. The UNSC has never legally mandated their creation meaning that individual states have responsibility for deciding how, and how quickly, to implement the WPS agenda limiting its impact on peacekeeping. In August 2022, 103 states had written a plan including most European countries.[40] It wasn't until UNSCR 2242 in 2015 that the UNSC 'welcomed' the creation of NAPs and called for other states to create them suggesting an acceleration of the agenda. Yet, concerns remain about the validity of NAPs. Aisling Swaine argues that nations with NAPs are largely donor countries looking to 'publicly demonstrate' commitment to gender mainstreaming.[41] The implication is that the impact is political and that there is limited real reform of peacekeeping missions. There is evidence to suggest reform is being slowed by national gender perspectives. In a 2015 open debate on WPS, the Russian delegation commented that NAPs 'cannot be used' to assess national policies and should only be 'voluntary'. Later in the debate, the Russian representative noted that women were facing 'increasingly'

37 Sabrina Karim, Kyle Beardsley, "Ladies Last", in *Gender, Peace and Security*, 68.

38 Georgina Holmes, "Situating Agency, Embodied Practices and Norm Implementing in Peacekeeping Training", *International Peacekeeping* 26,1 (2019).

39 Sabrina Karim, Kyle Beardsley, *Equal Opportunity Peacekeeping* (Oxford University Press, 2017).

40 For more information on NAPs, see: https://1325naps.peacewomen.org.

41 Aisling Swaine, "Globalising Women, Peace, and Security: Trends in National Action Plans" in *Rethinking National Action Plans*, 29.

serious threats underpinning a view of women as the victims of war and a masculine perspective of security.[42] In contrast, countries with NAPs have adopted different perspectives. For example, the UK NAP calls for further training and offers support to achieve the WPS agenda and redefine mission execution. Reflecting on Swaine's argument, this follows from the problems of 'celebrity diplomacy' insofar as the creation of a NAP offers a path to satisfy popular demands for progress but falls short of implementing real change.[43] Despite this, the expansion of NAPs should be considered a positive step in implementing the agenda. The difference in attitude between countries with and without NAPs, suggests that different regions of the world have been impacted differently by the WPS agenda and that the impact on peacekeeping has not been as deep or as uniform as intended.

Implementation and the British Army

The British Army has been conducting gender training for years, but often under different names. During operations in Iraq and Afghanistan the term Civil-Military Cooperation covered the basic functions expected by the WPS agenda. By looking through internal British Army archives, there are multiple online training courses available with the key word 'gender'.[44] However, there is evidence of a lack of gender training impacting operations. For example, a British Army Captain deployed with NATO in Kosovo critiqued the mission's intelligence structures for having a dominant masculine perspective which missed wider opportunities for the mission.[45] The officer considered that for intelligence collection to be successful, women were needed to talk to other women. This is particularly true in countries, such as Kosovo, which have been traumatised by sexual violence. She found that a lack of gender training and awareness in intelligence staff skewed analysis towards a masculine perspective, for example, when considering the impact of domestic violence. The mission predominately spoke with the male police commissioner who offered a male perspective, but rarely to women or victims because men could not access them, or did not think to, and lacked sufficient awareness. Domestic violence and the human trafficking of women is a significant issue in Kosovo, but the officer commented 'most

42 UNSC. Meeting Record S/PV.7533. New York, 13 October 2015.
43 Karen Eagle, "The Grip of Sexual Violence", in *Rethinking Peacekeeping*, 23.
44 Author search of British Army Knowledge Exchange, 21 July 22.
45 Author interview with British Army Captain deployed to Kosovo in 2021.

men just don't think these issues are relevant… this is a focus on fighters and extremists' reflecting a masculine perspective of human security. Similar deductions were made by the British Army on Op HERRICK in Afghanistan, with the campaign study noting that the mission was missing 50 per cent of the population.[46] The impact of not having sufficient knowledge of the WPS agenda can negatively affect operational effectiveness. Similarly, a lack of women soldiers to deploy in the first instance means policy changes are needed to recruit and retain more women.

The UK Ministry of Defence (MoD) has made significant efforts to improve human security training. The number of trained gender advisors is increasing. They receive a two-week course and further training once qualified to build knowledge and experience. Attendance at these courses is aimed at those who will deploy as gender advisors or in other specialist advisory posts. There is also a concerted effort to attend international courses and the MoD sponsors attendance on such international courses.[47] These training courses are aimed at individuals with a need to be trained and not at the wider force, however gender training is becoming more common.

Although the British Army is actively engaged in delivering human security training, there is less evidence of collective training in which groups of soldiers (instead of individuals) prepare for missions. A general weakness across both the British Army and the UN is an assumption that individual training courses will translate into mission effect. Such courses target selected individuals for training in the hope that they will influence the mission but does not test the skills a force may need. Closer to warfighting, examples of countering the Wagner Group in Mali provide a baseline of how to deal with the abuse of women and can be copied in training. A collective training event could see a force encounter a rape victim or witness domestic abuse for example. There is also recent operational experience to draw on. In Afghanistan, deployed forces' frequent encounters with the rape and abuse of young boys are documented in memoirs and the suppression of women was common. These scenarios pose ethical as well as operational questions about what to do, asking commanders to balance right and wrong against mission success. A wide gap in the implementation of the WPS agenda can be seen in the lack of collective training serials to prepare soldiers for such events in addition to being an opportunity to test doctrine and tactics.

46 MOD, HERRICK study, page 5-7_3.
47 MOD, JSP 1325 Part 2.

In speaking to officers, post-deployment, there are mixed opinions about the depth of impact the WPS agenda has had. One officer deployed to Kosovo in 2021 noted that a mission had two gender advisors but described them as being "decades behind UK thinking…some of the stuff the male advisor came up with was blatantly sexist". She also noted that some multinational staff were not always prepared to engage with women's groups, and this was reflected in their attitudes and gendered language towards them.[48] The mission did not engage with networks because of a mix of personal biases and an inability to view gender holistically. This is a single perspective in which we do not hear the context, but her commentary stands out and demonstrates the frustration many feel. In this case, it took the actions of one motivated individual with a high level of awareness to challenge the status quo. In another example, Major Tim Graham challenged the in-mission status quo to implement the WPS agenda in the Democratic Republic of Congo. His MBE citation reads his "sensitivity was widely apparent in his meticulous handling of allegations, both gender-based discrimination and incidents of child abuse".[49] There is a growing number of such cases in which British personnel actively implement the WPS agenda.

Perhaps the most significant area in which the British Army could change to better support women is with personnel policy. It was argued above that the Army struggles to recruit and retain sufficient women to meet the operational demand. The academic debate finds that social expectations of childcare and the need to maintain a work life balance are key factors preventing more women from serving.[50] The Armed Forces Continuous Attitude Survey has consistently found that the impact on personal life is the single biggest factor causing people to leave. In 2022, the survey found that 59 per cent of service leavers mentioned the impact of service life as the primary factor for them choosing to leave service, a trend this is consistent in reporting.[51] Meeting the WPS agendas demand for increased participation of women falls on a small percentage of the force, which in turn creates challenging force generation problems further exacerbating the impact on personal life. Despite high level political engagement, the Army

48 Author interview with British Army Captain deployed to Kosovo in 2021.
49 British Army, "2018 Honours and Awards" (2018), https://www.army.mod.uk/news-and-events/news/2018/11/operational-honours-and-awards-list-50-announced/.
50 Karim, Kyle, *Equal Opportunity Peacekeeping*.
51 MOD (2022), https://assets.publishing.service.gov.uk/government/uploads/system/uploads/attachment_data/file/1075579/Armed_Forces_Continuous_Attitude_Survey_2022_Main_Report.pdf.

has struggled to find meaningful changes to problems faced by women, demonstrated by the proportion of women choosing to serve remaining low. One example of political interest is former Minister Mark Francois who reported on how social changes impacted the Army.[52] Alongside highlighting childcare and the need for women-friendly policies, Francois explicitly called for more investment in people policies over equipment. That it has taken such interest to generate these recommendations, despite a pressing operational need, does imply that the Army is not yet ready to implement needed policy changes alone. More must be done to recruit and retain women. Policy making must specifically consider the importance of women and the operational advantage they bring. This alone would not solve all WPS issues, but it would help the British Army maintain the acceleration of the implementation of the participation pillar of the WPS agenda.

Institutionalisation

As an institution, the UN has been forward leaning with rhetoric reflecting the intent of UNSCR 1325. In 2003, the Secretary General laid out renewed procedures for enacting discipline in peacekeeping missions.[53] In 2005, the 'Zeid Report' also attempted to establish common rules for peacekeepers. Despite these efforts, the practical impact on peacekeeping missions was limited. Jasmine Kim-Westendorf used the 2014 example of the suppression of evidence of sexual exploitation and abuse as proof of institutional reluctance to punish and hold deployed soldiers accountable.[54] Elizabeth Defrais notes that all male deployments result in a 'wall of silence' which deters complainants and prevents peacekeepers from being punished effectively.[55] Indeed, Kim-Westendorf's evidence suggests that the lack of women deployed on missions is a critical weakness in breaking these barriers. However, it also suggests that predominately male staffed missions take divergent views on legal issues and whether behaviour is lawful or not. It also suggests that some contributing militaries are not yet

52 Mark Francois, "Stick or Twist" (2020), https://www.markfrancois.com/stick-or-twist.
53 UN Secretary-General, "Bulletin 9 October 2003", https://www.unhcr.org/uk/protection/operations/405ac6614/secretary-generals-bulletin-special-measures-protection-sexual-exploitation.html.
54 Jasmine Kim-Westendorf, "Abuse in Peacekeeping Operations" (2017), https://www.lse.ac.uk/women-peace-security/assets/documents/2017/wps9Westendorf.pdf.
55 Defeis, "U.N. Peacekeepers and Sexual Abuse".

prepared to engage with the WPS agenda. UNSCR 2272 'recalled' that the responsibility for investigating crimes committed by peacekeepers was on the troop contributing nations. This placed a large burden on contributing nations, which are often from developing regions of the world. Building on this, Gabrielle Simm identifies legal reasons undermining the intent of punishing crimes, for example, she cites difficulties of collating evidence for national legal systems and national concerns such as reputational damage.[56] As such, the problems are partly gendered, but also political and legal; they are broader than the attitude and culture of deployed missions. Whilst the WPS agenda has raised awareness, the lack of effective legal structures means it cannot be said to be fully institutionalised.

Since the passing of the UNSCR 1325, there have been sustained attempts to break the 'wall of silence' Defrais identified. In contrast to Yasushi Akashi's 1994 'boys will be boys' comment, the WPS literature has now grown significantly and internationally. This demonstrates a deeper and more coherent intellectual baseline challenging peacekeeping missions to evolve. Despite this growth, most authors writing on WPS are women and 'Western', suggesting that the academic penetration is limited. As theories of masculinity in security evolve, there is a growing recognition of the effect of toxic qualities of masculinity in operations including peacekeeping. A senior UK Civil Servant in the Ministry of Defence, Nick Pett, challenged men in defence to accept 'the existence of patriarchy' and that 'systems and processes' give men undue power and influence.[57] Looking at responses across social media to other articles calling for things such as gender-neutral language, a hard core remains who call such moves 'woke' and seem violently resistant to even engage with the arguments. Whilst Pett is a single voice it shows the debate has, in some nations, moved on from viewing women as victims of war and looks to them as gender equals. However, continued evidence of sexual abuse by peacekeeping forces suggests that its impact is predominately conceptual and academic. It will take longer for these attitudes to fully mature, but it shows an acceleration of the WPS agenda.

Pett's argument is reflective of a broader conceptual effect on the debate beginning to impact peacekeeping missions more directly. A 2019 report by SIPRI documented changes in the culture of the South Korean and Japanese militaries, as states with traditionally "patriarchal ideologies

---

56 Gabrielle Simm, *Sex in Peace Operations* (Cambridge, Cambridge University Press, 2013), 54.
57 Nick Pett, "So you want to be Feminists AF?", *Wavell Room* (2020), https://wavellroom.com/2020/04/23/so-you-want-to-be-feminist-af/.

and conservative gender norms".[58] Soumita Basu has also identified similar social trends in the Global South.[59] Engagement in this context appears to be bottom up and not directed by the UNSC. Social changes are proving to be more effective than institutional ones in creating impacts aligned to WPS and peacekeeping operations. The UNSC does not appear to share this understanding of the WPS agenda meaning they are unable to exploit it to reform peacekeeping. The formal agenda does not fully engage with the deeper social changes identified by Pett and SIPRI suggesting that it will take longer for practical changes to peacekeeping missions to be institutionalised.

Institutionalisation and the British Army

It is difficult to assess how these themes translate to the British Army. There is no evidence of the systematic cover up of crime and the Royal Military Police are largely capable of collecting credible evidence to prosecute where needed. To a large extent, the military justice system has already institutionalised the WPS agenda meaning the issues are cultural as well as legal and political. This has happened as British law has evolved and is incorporated into military institutions, alongside changing social expectations. The early adoption of a National Action Plan alongside a leading international role has given strategic leadership to the British Army.

However, there are worrying aspects to how the British Army operates and this indicates that more must be done to embed changes. In March 2012, it is alleged that a British soldier murdered Kenyan woman, Agnes Wanjiru.[60] The soldier reportedly stabbed her before dumping her body in a sceptic tank. Whilst reporting around the murder is inconclusive, it seems likely at the time of writing that the murder was an open secret with investigators met by a 'wall of silence'. Jasmine Kim-Westendorf suggests that predominately male missions take divergent views on moral issues and a reader may theorise if such a 'wall of silence' would have happened if a British or white woman who wasn't a sex worker had been murdered. Similar

58 Yeonju Jung and Ayako Tsujisaka, "Emerging Actors in the Women, Peace, and Security Agenda: South Korea and Japan" (2019), https://www.sipri.org/sites/default/files/2019-12/bp_1912_emerging_actors_in_the_wps_agenda.pdf.

59 Soumita Basu, "The Global South Writes 1324 (Too)", *International Political Science Review* 37, 3 (2016), 362-374.

60 Maya Oppenheim, "Kenyan Women Allegedly Murdered by British Soldier", *Independent*, 16 December 2021.

moments have been observed with 'Marine A' and the apparent murder of a Taliban fighter in Afghanistan suggesting that more must be done to train personnel to handle stressful situations. These are most likely isolated incidents in which justice will be found and not systematic problems; but the social conditioning around them suggests that more must be done to help understand right and wrong with better procedures for reporting.

These incidents represent the extreme of the WPS debate and not the reality of institutionalising the WPS agenda for the British Army. When discussing WPS, a civil servant with NATO experience in this field commented that "1325 needs to be everywhere all of the time or it is nowhere"[61] The evidence for the institutionalisation of 1325 is mixed against this criterion. In mission, there is often a general lack of awareness of how to institutionalise the needed changes. For example, reporting sexual harassment largely relies on trust in a local chain of command. However, there does not appear to be a standard practise to establish anonymous helplines or safe spaces. At unit level, the British Army relies heavily on religious figures to fulfil this role. The MoD has established a confidential helpline to act as an independent channel. The establishment of networks to study and represent women's issues has been successful but is largely focused on those impacted.[62] Soldiers anecdotally report different levels of support from their commanders suggesting that the role of networks is not yet established.

In the wider debate, the MoD has resisted calls to remove service legal primacy when dealing with accusations of sexual assault and rape. It was not until 2021, and the publication of the Atherton Report, that the Ministry agreed to deal with sexual complaints outside the chain of command. Sarah Atherton said "the fact that a servicewoman can now make a sexual complaint safe in the knowledge that her direct chain of command won't be handling it is a huge step forward".[63] She added that it was 'disappointing' that the MoD had refused to remove rape inquiries from the court martial system "despite clear evidence that the current system is failing to deliver justice".[64] It could also be argued that the British

61 Author interview with Civil Servant responsible for implementation of 1325. April 2022.
62 Kay Hallsworth, "Networks Within the Military", *Wavell Room* (2019) https://wavellroom.com/2019/07/11/networks-within-the-military-how-to-take-advantage-army-navy-airforce/.
63 Sarah Atherton quoted in "MoD overhauls rules for investigating sexual offences in armed forces", *Guardian*, 2 December 2021.
64 Dan Sabbagh, "MoD overhauls rules for investigating sexual offences in armed forces", *Guardian*, December 2, 2021.

Army is following wider social changes and not leading them. Of course, this analysis is focused on service culture and not the application of the WPS agenda when deployed on a mission. A glance at military journals such as the British Army Review and online journals such as the Wavell Room show that the average topics are still focus on strategy and policy. The editorial team of the Wavell Room commented, for example, "we were hoping for a bit more" after a call for papers surrounding the International Woman's Day 2021 #ChoosetoChallenge theme.[65] The lack of writing about how the British Army applies the WPS agenda, and a more open debate about strengths and weaknesses, suggests that the topic is still not embedded in military thinking.

Conclusion

This chapter has argued that the WPS agenda has impacted UN peacekeeping missions by challenging masculine theories and forcing conceptual and doctrinal changes. However, a level of strategic inertia remains across the spectrum of peacekeeping. The early stages of implementing the agenda were marked by simplistic measures such as counting women to judge success. As the UN deploys more peacekeeping missions, however, the mandates increasingly reflect the agenda's objectives with the UNSC responding to this challenge. The picture is mixed. The number of women deploying on peacekeeping missions has increased, but there remains an embedded theme that women only deploy on missions considered safe, which suggests that masculine perspectives still dominate national peacekeeping considerations.

But what do these themes mean? The impact of the WPS agenda on peacekeeping has been to challenge masculine doctrines and evolve concepts. The evidence shows an acceleration of change from 2008 but the impact remains limited with understandings fragile. International gender perspectives remain distanced from the practical implementation of peacekeeping with multiple views on how to evolve to meet future challenges. Despite evidence that embracing the WPS agenda improves the success of peacekeeping missions, the limited real impact over the past 20 years suggests that a core resistance remains. Masculine perspectives

65 "Choose to Challenge 2022 Update", *Wavell Room* (2022) https://wavellroom.com/â2022/01/19/choosetochallenge2022-women-in-defence/.

remain deeply embedded throughout the peacekeeping environment and within male dominated institutions best demonstrated by evidence of the deployment of women to 'safe' missions.

A more significant concern for supporters of the WPS agenda is that women are the majority stakeholder in this debate. William Hague noted that where men are 99 per cent of the problem, they must be at least 50 per cent of the solution.[66] The limited number of male writers, from both academia and within international militaries, suggests that the WPS agenda is having the greatest effect on its own academic understanding and within the political sphere instead of impacting peacekeeping missions directly. This is not to say that those involved in peacekeeping are ignorant of the agenda and its objectives. Rather, it suggests that masculine perspectives are deeply embedded, and more time is needed to challenge and replace them. Looking into the future, given the limited real impact on military peacekeeping it is also time to question if the WPS agenda has reached its full potential and what new tools might be needed to ensure gender parity in peacekeeping.

Looking back over 20 years, there was a level of inertia in WPS application. Evidence from operations since 2003 shows that the British Army is challenging and changing itself, demonstrating the impact of the agenda. There is certainly more that can be done, particularly with collective training, to fully embed the WPS agenda into the Army's operating model. The Army – and British society writ large – do not appear to have fully institutionalised the WPS agenda. Despite changes to the doctrinal and conceptual underpinning, including Joint Service Publications 1325[67] and 985[68], it is not at the forefront of military thinking and research showing it remains a specialised topic mirroring the academic literature.

It seems unlikely that this is because of embedded ideas about patriarchy. It is more likely that the case for how the WPS agenda improves operational performance has not yet been made convincingly. Military thinking prefers to think in terms of divisional tactics and campaigning

66 William Hague cited in Sebastian Whale, 'William Hague: "It is the work of a lifetime to comprehensively deal with sexual violence in conflict"', *The House*, 11 January 2019, https://www.politicshome.com/thehouse/article/william-hague-it-is-the-work-of-a-lifetime-to-comprehensively-deal-with-sexual-violence-in-conflict.

67 Ministry of Defence, Joint Service Publication 1325 Part 2: Guidance (2019), https://assets.publishing.service.gov.uk/government/uploads/system/uploads/attachment_data/file/770922/JSP1325_Part2.pdf.

68 Ministry of Defence, Joint Service Publication 985 (Human Security) (2021), https://www.gov.uk/government/publications/human-security-in-defence-jsp-985.

without the more complex understanding required when considering human security and WPS. Yet, as the British Army focuses on warfare, war evolves around human security and the importance of protecting civilians, and particularly women. In Afghanistan, the Army noted that it did not have access to 50 per cent of the population and this is arguably a key reason behind the operational failure. This mistake does not need to be made again. Understanding the WPS agenda is as vital to success on operations as knowing how to shoot.

Consolidating the deductions in this chapter, this chapter offers eight recommendations:

1. Publicly challenge stereotypical narratives about women in combat. The Army should lead in portraying women as soldiers and give them agency over their profession.
2. Challenge negative or stereotypical media representations of women in conflict and particularly combat; both from our force and how civilian women are portrayed.
3. Incorporate WPS serials into collective training. The Army should be as comfortable training around people as it is training for violence.
4. Identify prevailing social trends and amend personnel policies to reflect them. This chapter has specifically mentioned childcare and partners' careers. There will be others.
5. Incorporate more ethical training and discussions about right and wrong and social attitudes into training.
6. Incentivise writing about the WPS agenda and human security. This could be through the Defence Academy, with prizes, or linking scholarship to promotion prospects.
7. Increase efforts to recruit women into service.
8. Consider policies for the retention of women separate to those aimed at men.

These recommendations will allow the British Army to begin leading the WPS agenda and unlock its benefits, to direct its study, to train for its implementation, and to apply it in the field. Against the more general inertia evident in UN (and other) missions, this would give the British Army a chance to become a global leader in WPS implementation. What is really lacking for the British Army is a strong and concentrated desire to really 'do it'. The Army has many people actively engaged in the WPS agenda, but for the WPS agenda to be anywhere, it needs to be everywhere.

References

Barnes, Karen. "Reform of More of the Same?", YCISS Working Paper 41 (2006), https://genderandsecurity.org/sites/default/files/Barnes_-_Reform_or_more_of_the_same_0.pdf.

Basu, Soumita. "The Global South Writes 1324 (Too)", *International Political Science Review* (2016), 362-374.

British Army story boards from Mali 2022. UN Confidential document held by author.

Burton, James. "Cultural Apathy and Dishonesty within the British Army", *Wavell Room*, last modified 9 December 2020, https://wavellroom.com/2020/12/09/cultural-apathy-and-dishonesty-within-the-british-army-say-do-gap/.

British Army. "2018 Honours and Awards", last modified 23 November 2018, https://www.army.mod.uk/news-and-events/news/2018/11/operational-honours-and-awards-list-50-announced/.

Caparini, Marina. "Gender Training for Police Peacekeepers", last modified 4 October 2022, https://www.sipri.org/commentary/topical-backgrounder/2019/gender-training-police-peacekeepers-approaching-two-decades-united-nations-security-council.

de Jonge Oudraat, Chantal. "UNSCR 1325 – Conundrums and Opportunities", *International Interactions*, 39, 4 (2013), 612-619.

Defeis, Elizabeth. "U.N. Peacekeepers and Sexual Abuse," *Washington University Global Studies Law Review* 7, 2 (2008).

Eagle, Karen. "The Grip of Sexual Violence", in *Rethinking Peacekeeping, Gender Equality and Collective Security*, edited by Gina Heathcote & Dianne Otto. Basingstoke, Palgrave Macmillan, 2014.

Francois, Mark. "Stick or Twist", https://www.markfrancois.com/stick-or-twist.

Global Network of Women Peacebuilders. "Open Letter to Member States of the Security Council Re: Res 1960", last modified 20 October 2020 https://www.peacewomen.org/content/letter-open-letter-member-states-security-council-re-res-1960.

Hallsworth, Kay. "Networks Within the Military", *Wavell Room*, last modified 11 July 2019, https://wavellroom.com/2019/07/11/networks-within-the-military-how-to-take-advantage-army-navy-airforce/.

Heathcote, Gina. "Participation, Gender, and Security", in *Rethinking Peacekeeping, Gender Equality and Collective Security*, edited by Gina Heathcote & Dianne Otto. Basingstoke, Palgrave Macmillan, 2014.

Holmes, Georgina. "Situating Agency, Embodied Practices and Norm Implementing in Peacekeeping Training", *International Peacekeeping*, 26,1 (2019).

House of Commons Defence Committee, Protecting Those Who Protect Us: Women in the Armed Forces from Recruitment to Civilian Life, 25 July 2021, https://committees.parliament.uk/publications/6959/documents/72771/default/.

Hynes, Alan. "Progressive Insanities: Women in Defence", *Wavell Room*, last modified 20 April 2021, https://wavellroom.com/2021/04/20/progressive-insanities-women-in-defence/.

IPI roundtable, New York, 24 January 2020. "Profile: Yasushi Akashi", *Independent*, April 30, 1994.

Jung, Yeonju, Tsujisaka, Ayako. "Emerging Actors in the Women, Peace, and Security Agenda: South Korea and Japan", December 2019, https://www.sipri.org/sites/default/files/2019-12/bp_1912_emerging_actors_in_the_wps_agenda.pdf.

Karim, Sabrina, Beardsley, Kyle. "Ladies Last", in *Gender, Peace and Security*, edited by Theodora-Ismene Gizelis & Lousie Olsson Abingdon, Routledge, 2015.

Karim, Sabrina, Beardsley Kyle, *Equal Opportunity Peacekeeping*. Oxford University Press, 2017.

Kim-Westendorf, Jasmine. "Abuse in Peacekeeping Operations", September 2017, https://www.lse.ac.uk/women-peace-security/assets/documents/2017/wps9Westendorf.pdf.

Lloyd, Hugo. "British Army in Mali", *Wavell Room*, last modified 11 February 2022, https://wavellroom.com/2022/02/11/britisharmy-in-mali/.

Mackowski, Joanne. "Combat Inclusion: The End of Gender Segregation in the British Armed Forces", RUSI, last modified 11 July 2016, https://rusi.org/explore-our-research/publications/commentary/combat-inclusion-end-gender-segregation-british-armed-forces.

Ministry of Defence, Op HERRICK Campaign Study, https://assets.publishing.service.gov.uk/government/uploads/system/uploads/attachment_data/file/492757/20160107115638.pdf.

Ministry of Defence, Joint Service Publication 1325 Part 2: Guidance (2019), https://assets.publishing.service.gov.uk/government/uploads/system/uploads/attachment_data/file/770922/JSP1325_Part2.pdf.

Ministry of Defence, Joint Service Publication 985 (Human Security) (2021), https://www.gov.uk/government/publications/human-security-in-defence-jsp-985.

Olsson, Louise, Schjølse, Anita. Möller, Frida. "Women's participation in international operations and missions", in *Gender, Peace and Security*, edited by Theodora-Ismene Gizelis & Lousie Olsson Abingdon, Routledge, 2015.

Oppenheim, Maya. "Kenyan Women Allegedly Murdered by British Soldier", *Independent*, 16 December 2021.

Pett, Nick. "So You Want to Be Feminists AF?", *Wavell Room*, last modified 23 April 2020, https://wavellroom.com/2020/04/23/so-you-want-to-be-feminist-af/.

Sabbagh, Dan "MoD overhauls rules for investigating sexual offences in armed forces", Guardian, 2 December 2021.

Simm, Gabrielle. *Sex in Peace Operations*, Cambridge: Cambridge University Press, 2013.

Stiehm, Judith. "Women, Peacekeeping and Peacemaking" in *Women and International Peacekeeping*, edited by Louse Olsson & Torunn Tryggestad. London, Frank Cass, 2001.

Stone, Rosie. "The Third Sex", *Wavell Room*, last modified 29 January 2019, https://wavellroom.com/2019/01/29/the-3rd-sex-gender-and-contemporary-conflict/.

Tidblad-Lundholm, Kajsa. "When are Women Deployed?", *International Peacekeeping* 27, 4 (2020), 673-702.

UN, Report of the Economic and Social Council for the Year 1997, A/52/3/Rev.1.

UN, UN Peacekeeping Operations Principles and Guidelines, (2008) https://www.un.org/ruleoflaw/files/Capstone_Doctrine_ENG.pdf accessed 21 July 2022.

UN, "Ten Year Impact Study on UNSCR 1325", accessed 20 October 2020, https://www.globalpolicy.org/images/pdfs/10year_impact_study_1325.pdf.

UN Secretary-General, "Bulletin 9 October 2003", last modified 9 October 2003, https://www.unhcr.org/uk/protection/operations/405ac6614/secretary-generals-bulletin-special-measures-protection-sexual-exploitation.html.

UN Women, "Preventing Conflict, Transforming Justice, Securing Peace", http://peacewomen.org/sites/default/files/UNW-GLOBAL-STUDY-1325-2015%20(1).pdf.

UK Government, National Security Strategy and Strategic and Security Review 2015, https://www.gov.uk/government/publications/national-security-strategy-and-strategic-defence-and-security-review-2015.

UNSC. Meeting Record S/PV.7533. New York, 13 October 2015.

Wavell Room. "Choose to Challenge 2022 Update", *Wavell Room*, last modified 19 January 2022, https://wavellroom.com/2022/01/19/choosetochallenge2022-women-in-defence/.

Woodward, Rachael and Winter, Patricia. "Discourses of Gender in the Contemporary British Army." *Armed Forces & Society* 30, 2 (2004): 279–301.

4

HEARTS, MINDS, WOMEN, AND PEACE

US Militarized Human Security and Its Discontents

Kelly Atkinson

For twenty years following the attacks on September 11th 2001, "winning hearts and minds" circulated as common parlance among military leaders, policymakers, and households during the United States' sustained Global War on Terror (GWOT). Developed by General David Petraeus and codified into policy through the 2006 Army Counterinsurgency manual,[1] the concept of winning hearts and minds signified the United State (US) military's focus on human security in the twenty-first century. With the publication of the U.S. Department of Defense's (DoD) "Women, Peace and Security Strategic Framework and Implementation Plan" in June 2020,[2] a human security focus now functions as an integral element of US military engagements. Yet with the collapse of the Afghan government following US military withdrawal in August 2021, all signs indicate that the winning hearts and minds policy ultimately failed. Given this outcome, why might we anticipate that the US military could achieve success implementing another human security policy approach where they failed before?

The policies of "winning hearts and minds" and "women, peace and security" both represent US military operationalization of human security,

1 Manual, Field. "Manual 3-24: Counterinsurgency." *Department of the Army, Washington DC*, 2006.

2 U.S. Department of Defense. "Women, peace, and security strategic framework and implementation plan." (2020). Available at https://media.defense.gov/2020/Jun/11/2002314428/-1/-1/1/WOMEN_PEACE_SECURITY_STRATEGIC_FRAMEWORK_IMPLEMENTATION_PLAN.PDF.

or "the security of individuals and communities rather than the security of states."[3] Despite falling under the same security category, these policies differ in distinct and important ways: winning hearts and minds functioned as a transactional policy, whereas the Women, Peace and Security (WPS) agenda offers opportunities for transformational policy engagement. This key policy element provides potential for WPS to succeed where winning hearts and minds failed. This chapter argues that in order to capitalize on this potential for success, the US military must understand the dynamics of human security policy goals, implementation, outcomes, and challenges.

A transactional policy refers to one in which an actor – in this case, the US military – maximizes power in the relationship and executes policy to achieve a desired outcome. Hearts and minds operated in such a way since the US military developed and applied this policy to execute counterinsurgency operations in Afghanistan and Iraq over a twenty-year period. Power flowed in a unidirectional manner, with an occupying military force seeking to impose its will upon subject populations. Under such a transactional policy, the relationship ends when the powerful actor achieves its goals (or, in this case, fails to do so). There is no expectation for change or adjustment on behalf of the imposer since the actor holding power controls the policy relationship. In sum, the US policy of winning hearts and minds represents the key elements of transactional foreign policy: it "focuse[d] on short-term wins rather than longer-term strategic foresight, adhere[d] to a zero-sum worldview where all gains [were] relative and reciprocity [was] absent, reject[ed] value-based policymaking, and [did] not follow a grand strategy."[4]

Alternatively, WPS policy offers transformational opportunities since the approach involves "paradigmatic shifts, the reconfiguration of interests, [and] large scale institutional re-ordering and changed logics."[5] Unlike the unidirectional nature of winning hearts and minds, WPS demands that the actor imposing this policy – the US military – conducts internal institutional change alongside pursuing external policy engagements. Specifically, Defense Objective 1 under the US WPS strategic plan requires that the "Department of Defense exemplifies a diverse organization that

3 Kaldor, Mary. *Human Security*. Polity, 2007, 182.

4 Bashirov, Galib, and Ihsan Yilmaz. "The rise of transactionalism in international relations: Evidence from Turkey's relations with the European Union." *Australian Journal of International Affairs* 74, no. 2 (2020): 165.

5 Ashbee, Edward and Steven Hurst. "The Trump foreign policy record and the concept of transformational change." *Global Affairs* 6, no. 1 (2020): 5-19.

allows for women's meaningful participation across the development, management, and employment of the Joint Force."[6] This stipulation ensures that the US military transforms its own institutional make-up while also engaging with external partners, which offers opportunities for successful policy engagement of militarized human security. Indeed, this chapter argues that Defense Objective 1 is the key policy element which distinguishes the WPS agenda from the winning hearts and minds approach.

This chapter proceeds as follows: first, it explores human security within the dynamics of military engagements. The chapter examines how the US military operationalized human security through its policy of winning hearts and minds in the counterinsurgency conflict in Iraq and Afghanistan from 2001 to 2021. Next, the chapter explores the emergence of the WPS policy agenda and traces how the US recently adopted this framework. Through comparing the transactional nature of winning hearts and minds with the transformational potential of WPS, the chapter identifies opportunities for the US military to employ successful human security engagements. The chapter concludes with recommendations regarding how the US and other powers must adopt a reflective, self-critical policy perspective in order to capitalize on the opportunities for human security application within military engagements.

Militarized Human Security

While this chapter explores the relatively novel human security focus within US military operations over the past two decades, human security itself is not a new phenomenon. With the creation of the United Nations (UN) in 1945 and the subsequent establishment of UN organizations focused on specific mission sets, human security policy is foundational to global governance in the twentieth century. The United Nations Development Programme's *Human Development Report 1994: New Dimensions of Human Security* codified the UN's prior decades of emerging focus on human security by "equating security with people rather than territories, with development rather than arms."[7] Today, a well-established body of literature reveals the connections between issues of human security and conflict, from the cooptation of

6 U.S. Department of Defense. "Women, peace, and security strategic framework and implementation plan." (2020), 7.

7 United Nations Development Programme. *Human Development Report 1994: New Dimensions of Human Security*, New York (1994).

humanitarian aid by terrorist organizations[8] to the potential for empowered women to protect displaced children from recruitment into armed groups.[9]

Against this policy backdrop, this chapter explores **militarized human security** defined as the implementation of an individual and community security focus within military doctrine with the goal of improving operational mission effectiveness. Given the military's function as the continuation of politics by other (violent) means,[10] executed by officers who themselves serve as the managers of state violence,[11] it may seem antithetical to imagine military doctrine considering the security of individuals and communities as worthy of attention. Indeed, with the US military's colloquial mission to 'kill people and break things', the individuals and communities involved are often those killed or broken. Yet with the (temporary) cessation of great power conflict at the end of the Cold War and the burgeoning impacts of economic and social globalization at the turn of the century, the US military's pivot to engaging the security of individuals and communities as foundational to achieving its counterinsurgency mission emerged as a clear evolution of defense policy.

"Winning Hearts and Minds"

The US military first developed an intentional, overt human security focus within its operational policies during the twenty-year GWOT. After the 9/11 attacks, America launched a retaliatory war against Afghanistan followed shortly thereafter by its invasion of Iraq in 2003. In both campaigns, the US faced non-standard adversaries in uneven terrain: ousting the Taliban regime in Afghanistan quickly devolved into sustained coalition operations against locally embedded tribal factions, and sectarian divides within Iraqi populations generated a similarly complex web of shifting allies and adversaries for US troops in the country. US senior military leadership faced this complex era of warfare by incorporating an evolving approach to operational engagements: a human security approach.

8 Choi, Seung-Whan, and Idean Salehyan. "No good deed goes unpunished: refugees, humanitarian aid, and terrorism." *Conflict Management and Peace Science* 30, no. 1 (2013): 53-75.

9 Atkinson, Kelly E. "Displacement, gender-based development, and grave violations against children in armed conflict: the case of Iraq." *Journal of Peacebuilding & Development* 17, no. 1 (2022), 108-126.

10 Von Clausewitz, Carl. "On war." In *On War*. Princeton University Press, 2008.

11 Huntington, Samuel P. *The soldier and the state: The theory and politics of civil–military relations.* Harvard University Press, 1981.

Policy never emerges from a vacuum. In the post-9/11 phase of human security engagement, the scholar-practitioner who shaped US military policy more than any other was General David Petraeus.[12] For his 1987 PhD from Princeton's Woodrow Wilson School of International Affairs, then-Major Petraeus explored "The American Military and the Lessons of Vietnam: A Study of Military Influence and the Use of Force in the Post-Vietnam Era."[13] In considering the "impact of Vietnam on America's senior military with respect to their most important task – advising the nation's leadership on the use of American military forces in potential combat situations,"[14] Petraeus concluded that "caution...is likely to characterize the military approach to the use of force for some time."[15] Arguing against absolute caution derived exclusively from a stark adherence to lessons learned, Petraeus assessed that:

> American involvement in counterinsurgencies is almost universally regarded as more likely than involvement in most other types of combat... the senior military is thus in a dilemma. The lessons taken from Vietnam would indicate that, in general, involvement in a counterinsurgency should be avoided. But prudent preparation for a likely contingency (and a general inclination against limiting a president's options) lead the military to recognize that significant emphasis should be given to counterinsurgency forces, equipment, and doctrine.[16]

Forming this mindset while a field grade officer assigned as an Assistant Professor in the Department of Social Sciences at the U.S. Military Academy, Petraeus soon applied his lessons learned from Vietnam when writing the Army's *Field Manual 3-24 Counterinsurgency*, published in 2006 after five years of counterinsurgency operations in Afghanistan. Writing as a Lieutenant General in command of the U.S. Army Combined Arms Center, Petraeus and co-author Lieutenant General James F. Amos, U.S. Marine Corps, produced a manual "designed to fill a doctrinal gap... with our Soldiers and Marines fighting insurgents in Afghanistan and

---

12 This chapter focuses on General Petraeus as the origin of the US military's "winning hearts and minds" focus since he not only developed the theory academically but implemented it as a strategic practitioner.

13 Petraeus, David Howell. *The American military and the lessons of Vietnam: a study of military influence and the use of force in the post-Vietnam era*. Princeton University, 1987.

14 Ibid., i.

15 Ibid., ii.

16 Petraeus, David H. "Lessons of history and lessons of Vietnam." *The US Army War College Quarterly: Parameters* 16, no. 1 (1986): 27, 49.

Iraq, it is essential that we give them a manual that provides principles and guidelines for counterinsurgency operations. Such guidelines must be grounded in historical studies. However, it also must be informed by contemporary experiences."[17]

Assessing the purpose and structure of the manual, Vizzard and Capron argue that:

> The manual's publication marked a major departure from military practice, both in the way in which it was developed and in the doctrine it espoused. It attempted to meld the hard-won lessons of ground commanders in Iraq with the insights of academics who had studied insurgencies and counterinsurgencies to produce a functioning doctrine that would help the United States reverse the rapidly deteriorating Iraq situation...the manual appropriately places political action at the forefront of counterinsurgency and subordinates military operations – particularly offensive operations – to the establishment of functioning, legitimate government institutions. However, the publication of such a manual by the military, and particularly by the two ground components of the military, rather than by the Department of Defense at the joint level, highlights the manual's inadequacy for achieving its stated purpose.[18]

With *Field Manual 3-24,* Petraeus synthesized his doctoral studies of lessons learned from the US failure in Vietnam with his own assessments regarding the future of counterinsurgency operations. In doing so, he produced the doctrine which shaped the US military's GWOT campaign in post-9/11 Middle East. Thus, we see the origins of the "hearts and minds" era of militarized human security embedded within the doctrine.[19]

Specifically, *Field Manual 3-24* provides the first codified example of the US military implementing an individual and community security focus within military doctrine with the goal of improving operational mission effectiveness. For example, in the manual's Appendix A, "A Guide for Action," we find instructions on "translating lessons of this manual into practice."[20] Instructions for executing counterinsurgency plans include a

17 "Manual 3-24: Counterinsurgency," Foreword.

18 Vizzard, James W., and Timothy A. Capron. "Exporting General Petraeus's counterinsurgency doctrine: an assessment of the adequacy of Field Manual 3-24 and the US Government's implementation." *Public Administration Review* 70, no. 3 (2010): 485-486.

19 The phrase "hearts and minds" originates from Sir Gerald Templar's association of the phrase with the British counterinsurgency campaign in Malaya, 1948-1960. See Dixon, Paul. "'Hearts and minds'? British counter-insurgency from Malaya to Iraq." *Journal of Strategic Studies* 32, no. 3 (2009): 353-381.

20 Manual, Field. "Manual 3-24: Counterinsurgency." A-1.

section on building trusted networks, a cornerstone of counterinsurgency (COIN) operations:

> This is the true meaning of the phrase "hearts and minds," which comprises two separate components. "Hearts" means persuading people that their best interests are served by COIN success. "Minds" means convincing them that the force can protect them and that resisting it is pointless. Note that neither concerns whether people like Soldiers and Marines. Calculated self-interest, not emotion, is what counts. Over time, successful trusted networks grow like roots into the populace. They displace enemy networks, which forces enemies into the open, letting military forces seize the initiative and destroy the insurgents.[21]

As ubiquitous as the phrase "hearts and minds" became during the US's protracted wars on terrorism, the aforementioned paragraph represents its singular mention within the 282 pages of the counterinsurgency manual. Yet its impact was profound, serving as a recognizable talking point among senior military and defense officials during the GWOT years.[22] Indeed, the language spans party ideologies. President George W. Bush justified the invasion of Iraq to the UN General Assembly in 2005 by saying "across the world, hearts and minds are opening to the message of human liberty as never before." [23] President Barack Obama later employed the language in an attempt to rebuild relations with the Muslim world in 2009 by asserting that abiding by Geneva Conventions "will make us safer and will help in changing hearts and minds in our struggle against extremists."[24]

The policy also garnered considerable critique from practitioners and academics alike, from the asymmetric application of COIN operations[25] to the normative assumptions of modernization and legitimacy tied to Westernizing local populations through this militarized human security approach.[26] Ultimately, with the collapse of the Afghan government following the US military withdrawal from the country in August 2021,

21 Ibid., A-5.

22 Petraeus, David H. *Learning counterinsurgency: Observations from soldiering in Iraq*. Army Combined Arms Center, Fort Leavenworth KS (2006).

23 Dickinson, Elizabeth. "A bright shining slogan: how 'hearts and minds' came to be." *Foreign Policy*, 22 August 2009. Accessible at https://foreignpolicy.com/2009/08/22/a-bright-shining-slogan/.

24 Ibid.

25 Nell, Karl E. "Hearts-and-minds: a strategy of conciliation, coercion, or commitment?" Army War College, Carlisle Barracks PA (2012).

26 Egnell, Robert. "Winning 'hearts and minds'? A critical analysis of counter-insurgency operations in Afghanistan." *Civil Wars* 12, no. 3 (2010): 282-303.

many now assess that the policy to win hearts and minds failed in the end.[27]

Turning to doctrine offers possible explanations for this policy failure. At its core, winning hearts and minds is a *transactional* policy built upon systems of reward and discipline.[28] According to *Field Manual 3-24*, winning hearts and minds relies upon persuasion, convincing, and calculated self-interest. In this framework, operationalizing human security becomes a project of imposing US national interests through persuasion that often becomes forceful in execution. For example, policies of "enhanced interrogation" and "extraordinary rendition" imposed under the framework of winning hearts and minds illustrate the forceful nature of this policy.[29] Whether through these examples of force or through more benign examples of persuasive relationship-building, as in implementation of female engagement teams in Afghanistan,[30] the goal of building trusted networks ultimately benefits the imposer and not the subject, with minimal enduring concern regarding the interests of the individuals and communities involved. The doctrine frames any change occurring during the engagement between military imposer and subject as unidirectional: whether local populations like Soldiers and Marines is inconsequential, since the U.S. military holds the position of power. This unidirectional flow of power reinforces the transactional nature and consequent limitations of winning hearts and minds.

Furthermore, militarized human security emerged during the era of mission creep for the US military.[31] Winning hearts and minds as counterinsurgency policy entailed elements of nation-building and democratization spanning an undefined time horizon, with the executors of such engagement falling under the DoD rather than the State Department. Militarizing US foreign policy in this way complicated the purpose and mission of winning hearts and minds during the GWOT years: did the US military incorporate a human security perspective to defeat insurgents in Iraq and Afghanistan? Or did the US military integrate human security

---

27 Hazelton, Jacqueline L. "The hearts-and-minds myth: how America gets counterinsurgency wrong." *Foreign Affairs*, 15 July 2021.

28 Bass, Bernard M., and Ronald E. Riggio. *Transformational leadership*. Psychology press, 2006.

29 Mayer, Jane. *The Dark Side: The Inside Story of How the War on Terror Turned into a War on American Ideals*. Anchor, 2009.

30 Keohane, Jennifer and Kelly Jakes. "Soldiers and scholars: evaluating female engagement teams in the war in Afghanistan." *Women's Studies in Communication* 44, no. 1 (2021): 102-118.

31 Adams, Gordon, and Shoon Murray, eds. *Mission creep: the militarization of US foreign policy?* Georgetown University Press, 2014.

within its doctrine in order to democratize these countries? Winning hearts and minds spanned both approaches, resulting in an overextended military incapable of completing either task effectively.

In the end, the US military employed a militarized human security approach to engage individuals and communities in Afghanistan and Iraq to find, fix, and disrupt terrorist operations and insurgencies. Yet winning hearts and minds operated as a transactional policy approach imposed through force without a clearly delineated purpose. For these reasons, the US's first doctrinal instance of militarized human security failed.

As two decades of GWOT operations come to a close with the US withdrawal from Afghanistan, a new era of militarized human security policy is emerging with the Department of Defense's implementation of the Women, Peace and Security policy agenda.[32] Given the failure of winning hearts and minds over the last twenty years, why would we expect the US to achieve success implementing another human security policy approach where they failed before?[33]

The Women, Peace and Security Policy Agenda

The WPS agenda originated in 2000 with the United Nations Security Council Resolution (UNSCR) 1325 on women, peace and security.[34] This policy framework called for the United Nations and member states to serve as global leaders in promoting the meaningful participation of women in conflict prevention, management, and resolution. Specifically, the WPS agenda "outlines actions to be taken by the Secretary General, the Security Council, UN departments and member states to 'mainstream gender' into peace and security policies and practices."[35] Originating at the UN, the

32 U.S. Department of Defense. "Women, Peace, and Security strategic framework and implementation plan." (2020).

33 Some scholars and policymakers are hopeful that twenty years of foreign engagement has set a foundation for democratic leanings in Afghanistan and Iraq, and the author certainly shares their optimism. But, for the purpose of this chapter, analysis focuses exclusively on US military operations and the implementation of militarized human security. The US military failed to achieve its objectives in Iraq (somewhat) and Afghanistan (fully), so this chapter argues that the winning hearts and minds policy failed.

34 United Nations Security Council Resolution 1325, "Women, peace and security." S/RES/1325 (2000). Available at https://documents-dds-ny.un.org/doc/UNDOC/GEN/N00/720/18/PDF/N0072018.pdf?OpenElement.

35 Cohn, Carol, Helen Kinsella, and Sheri Gibbings. "Women, peace and security resolution 1325." *International Feminist Journal of Politics* 6, no. 1 (2004): 130.

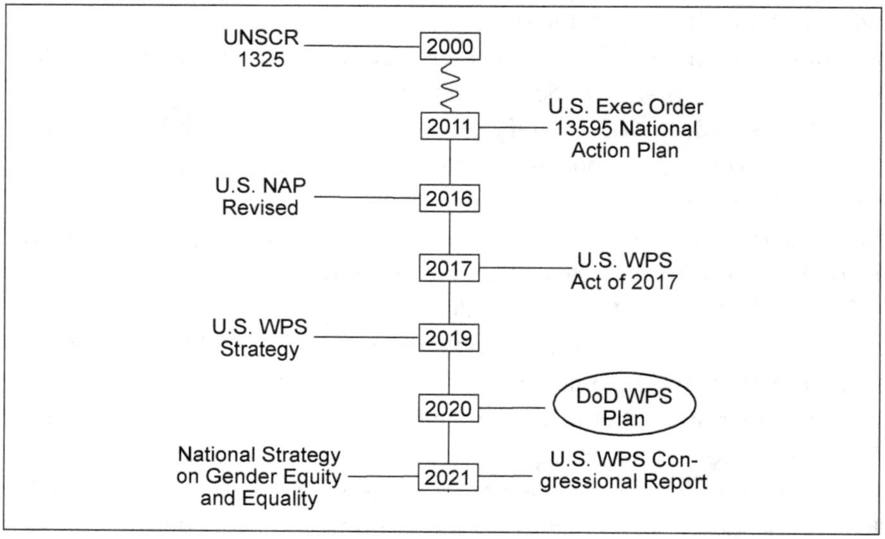

Figure 4.1. US WPS Timeline

WPS policy agenda achieves global diffusion[36] through state-level National Action Plans that dictate how governments will implement WPS principles and tenets domestically and through foreign policy engagements.

The US arrived late to the WPS game implementing its first National Action Plan via executive order eleven years after the adoption of UNSCR 1325 and releasing a comprehensive plan five years later. In quick succession, the U.S. Congress passed the Women, Peace, and Security Act of 2017 followed by the U.S. Strategy on Women, Peace, and Security in 2019. This 2019 Strategy supersedes previous National Action Plans and initiated department-level frameworks, to include the DoD Women, Peace, and Security Strategic Framework and Implementation Plan published in June 2020. Figure 4.1 summarizes this timeline.

Within the DoD WPS Plan, we can trace the next iteration of US militarized human security policy. The DoD Plan nests under the aforementioned 2019 Strategy and its four Lines of Effort (LOE):

> (1) seek and support the preparation and meaningful participation of women around the world in decision-making processes related to conflict and crises; (2) promote the protection of women and girls' human rights; access to humanitarian assistance; and safety from violence, abuse, and

36 True, Jacqui. "Explaining the global diffusion of the women, peace and security agenda." *International Political Science Review* 37, no. 3 (2016): 307-323.

exploitation around the world; (3) adjust U.S. international programs to improve outcomes in equality for, and the empowerment of, women; and (4) encourage partner governments to adopt policies, plans, and capacity to improve the meaningful participation of women in processes connected to peace and security and decision-making institutions.[37]

The DoD developed its WPS Plan to detail the department's roles and responsibilities for implementing the 2019 Strategy and established WPS Defense Objectives to support the Lines of Effort. These specific Defense Objectives "provide the Department with long-term end states to guide the Department's implementation of the [2019 Strategy] over time:"[38]

- Defense Objective 1. The Department of Defense exemplifies a diverse organization that allows for women's meaningful participation across the development, management, and employment of the Joint Force.
- Defense Objective 2. Women in partner nations meaningfully participate and serve at all ranks and in all occupations in defense and security sectors.
- Defense Objective 3. Partner nation defense and security sectors ensure women and girls are safe and secure and that their human rights are protected, especially during conflict and crisis.[39]

Additionally, the plan aims to support the U.S. National Security Strategy which outlines that "societies are more peaceful and prosperous when women and men enjoy the same rights, liberties, dignities, and access to resources."[40] As such, the DoD sets goals within its WPS Plan to:

- **Build a more lethal force** by providing the tools necessary to reduce operational risk in a multi-domain environment, and recruit and fully leverage a diverse and innovative fighting force;
- **Strengthen alliances and attract new partners** by demonstrating U.S. commitment to human rights and women's empowerment, making the United States the partner of choice; and

37 U.S. Department of Defense. "Women, Peace, and Security strategic framework and implementation plan." (2020), 7.
38 Ibid., 8.
39 Ibid., 7.
40 Ibid., 10.

GLOBAL	NATIONAL	DEPARTMENTAL
WPS Principles • Participation of women in peace and security • Protection of women and girls from violence • Inclusion of women in conflict prevention • Equal access to relief and recovery before, during, and after conflict and crisis • Protection of human rights • Equal application of the rule of law • Incorporation of a gender perspective into peace and security efforts	*WPS Strategy* LOE 1 directs the Department to: • Seek and support women's meaningful participation in military decision-making processes, disaster preparation and response, and stabilization. • Increase the meaningful participation of women in partner nation security sector initiatives, including programs on the rule of law and within professional military education. • Lead by example through inclusion of American women in U.S. efforts abroad. • Leverage relevant analysis indicators, including the collection of sex-and-age-disaggregated data, to identify and address barriers to women's meaningful participation.	DoD Equities Supporting WPS Principles • Diversity & Inclusion • Gender Integration • Inclusive Leadership Development • Professionalization of Partner Nation Armed Forces • Recruitment & Retention • Secual Harassment & Assault Prevention • Sexual Exploitation &Abuse Prevention • Gender-Based Violence Prevention • Protection of Civilians • Protection of Children Affected by Armed Conflict • Countering Trafficking in Persons • Humanitarian Assistance & Disaster Relief • Countering Violent Extremist Organizations • International Humanitarian Law • International Human Rights Law • Protection of Cultural Property
	WPS Strategy LOE 2 directs the Department to: • Promote the protection of women and girls' security, human rights, and access to aid with governments and regional or other security sector forces as appropriate. • Address security-related barriers to the protection of women and girls. • Prioritize efforts to prevent and respond to sexual exploitation and abuse. • Provide women and girls with safe and equal access to humanitarian assistance. • Empower women as partners in preventing and combatting terrorism.	
	WPS Strategy LOE 3 directs the Department to: • Adjust its international programs to improve outcomes in women's equality and empowerment. • Train DoD personnel on the needs, perspectives, and security requirements of men and women; protecting civilians from violence, exploitation, and trafficking in persons; and international humanitarian law (IHL) and international human rights law (IHRL). • Apply gender analyses to improve DoD program design and targeting.	
	WPS Strategy LOE 4 directs the Department to: • Encourage partner nation governments to adopt policies, plans, and capacity to improve the meaningful participation of women in processes connected to security and decision-making institutions. • Work with partner nations to remove legal, regulatory, and structural barriers faced by women in defense and security sectors. • Assist partner nations in increasing opportunities for women to serve in security sector forces, including peacekeeping and military organizations, by developing their technical and professional competencies.	

Table 4.1. Women, Peace, and Security Principles, National Lines of Effort, and DoD Equities.

Source: U.S. Department of Defense. "Women, Peace, and Security strategic framework and implementation plan" (2020).

- **Reform the Department for greater performance and affordability** by developing more effective strategies to mitigate risks and optimize mission success.[41]

The DoD's WPS Plan is comprehensive in scope, linking the overarching WPS principles originating in UNSCR 1325 to the strategic policy level and operational impacts of the US military service branches. Table 4.1 depicts the various existing DoD equities that align within the WPS framework, which opens doors for DoD leaders and organizations to incorporate WPS policies within existing efforts. Additionally, the 2021 and 2022 U.S. Government Women, Peace, and Security Congressional Reports outline the mechanisms through which the DoD implements the WPS agenda, identifying the steps taken to implement the DoD's WPS Plan.

Examining these steps reveals the internal mechanisms through which the DoD engages this nascent phase of militarized human security, in which women and girls comprise the individuals and communities subject to the approach. As documented in the 2021 and 2022 Congressional WPS reports, DoD highlights include training and embedding full-time Gender Advisors within Combatant Commands, developing Gender Focal Point staff networks to coordinate WPS implementation, funding activities designed to build partner capacity to advance WPS principles, and incorporating WPS principles within existing training and education programs for DoD personnel.[42]

The DoD has taken measures to translate WPS policy into doctrine, revising Army Field Manuals and "numerous Joint Publications on topics including intelligence, operations, security cooperation, and planning, among others."[43] Finally, the 2022 Congressional report specifically addresses how the DoD accomplished internal transformations to achieve WPS objectives: "coordination on WPS with offices that oversee the Department's diversity, equity, inclusion, and accessibility and sexual harassment and assault prevention and response was particularly important in developing implementation plans following the review conducted by the Independent Review Commission on Sexual Assault in the Military, initiated in February 2021."[44]

41 Ibid., 10.
42 The White House. "USG women, peace, and security congressional report." 29 June 2021, A-34.
43 Ibid., A-35.
44 The White House. "USG women, peace, and security congressional report." July 2022, 43.

Policy serves as the foundation of mission execution, operationalized through doctrine. The DoD's incorporation of WPS policy into joint doctrine is critical to implementing the established WPS agenda within US military operations, just as the Army *Field Manual 3-24* operationalized winning hearts and minds over twenty years of COIN engagements. Operationalized outcomes of winning hearts and minds during the GWOT decades range from the female engagement teams operating in Iraq and Afghanistan to militarized efforts to build schools and hospitals in local communities.[45] At its nascent stage, the US military's WPS agenda has not yet produced robust operationalized outcomes – the doctrinal foundation is established, and the operational impact is newly emerging through DoD internal processes and external engagements. Thus, we stand within a liminal space in which the potential for WPS to produce successful policy outcomes still exists, whereas winning hearts and minds ultimately failed. As such, we must interrogate how and why the WPS agenda differs from its militarized human security predecessor.

Comparing Policies: Why is WPS Different?

As evident across prolific policy documents, the DoD maintains a comprehensive strategic framework through which to establish and execute WPS principles in order to serve "as a global model for diversity and inclusivity while working with partner nations to support women's meaningful participation within the defense and security sectors."[46] WPS resides firmly within the realm of human security, and DoD implementation of WPS principles represents the contemporary era of US militarized human security. While the WPS agenda is well-established, DoD operationalization is recent and follows two decades of militarized human security policy that ultimately failed to achieve its objectives. Why might human security through WPS succeed where winning hearts and minds failed?

The potential for a successful WPS human security approach within the US military lies in three mechanisms: WPS entails 1) a transformational policy approach that is 2) relational and participatory with 3) clear mission

45 Fishstein, Paul and Andrew Wilder. "Winning hearts and minds? Examining the relationship between aid and security in Afghanistan." *Feinstein International Center*, Tufts University, Medford, MA 50 (2012).

46 U.S. Department of Defense. "Women, Peace, and Security strategic framework and implementation plan." (2020), 8.

	Hearts and Minds	Women, Peace and Security
Purpose	1. Defeat insurgencies in Iraq and Afghanistan 2. Nation-build? 3. Democratize?	1. Diversify Department of Defense 2. Support women's defense/ security participation in partner nations 3. Protect women/girls in conflict
Ways	Transactional	Transformation
Means	Impose through force	Participatory and reflective
Ends	Outright Policy Failure	To be determined

Table 4.2. Comparison of Militarized Human Security Policies.

objectives. These elements distinguish the agenda from winning hearts and minds, which was transactional and imposed through force without clear metrics of success. The differences are key and offer opportunities for US military leaders to succeed in executing militarized human security, defined previously as *the implementation of an individual and community security focus within military doctrine with the goal of improving operational mission effectiveness*.

Compared to the *transactional* nature of winning hearts and minds, the DoD's implementation of the WPS agenda is *transformational*. The DoD WPS policy aims not only to work with partner nations to promote and advance gender inclusion locally and within partner military forces but also to transform the institutions of the DoD itself. As discussed earlier, Defense Objective 1 directs the DoD to exemplify "a diverse organization that allows for women's meaningful participation across the development, management, and employment of the Joint Force."[47] This approach turns the lens of examination inward and demands that the DoD confronts its own internal practices preventing gender equality, from systematic barriers to diverse recruitment and retention[48] to pervasive practices of gender-based violence in the military.[49]

For example, the DoD's focus on training Gender Advisors to educate the services and implement WPS principles at the Combatant Command level reflects the participatory nature of this militarized human security

47 The White House. "USG women, peace, and security congressional report." 29 June 2021, A-34.

48 Atkinson, Kelly and Alea Nadeem. "Warrior braids and the Air Force women's initiative team – the invisible labor behind diversity, inclusion, and institutional change." *Wild Blue Yonder*, 17 May 2021.

49 Hunter, Kyleanne and Emma Moore. "Implementing women, peace, and security guidance to systematically combat gender-based violence in the military." *A Policy Brief from the Athena Leadership Project*, 11 May 2021.

framework, since all military members must engage with gender equality training and education. Moreover, the DoD is moving towards diversifying its education and training practices more broadly, from promoting diversity, equity and inclusion within military service academies[50] to updating annual training requirements in order to remove stereotypes and cultural biases.[51] These key changes embody the internal reflection required to achieve the DoD's goal of serving "as a global model for diversity and inclusivity," as codified in the WPS policy framework.[52]

Additionally, the DoD cannot urge partner nations to recruit more women into the military or send more women to professional military education opportunities if its own service branches fail to do so. As a result, Defense Objective 1 within the WPS framework holds the DoD accountable for implementing gender equality principles internally in order to extol their virtues externally through relational engagements, thus enabling the accomplishment of Defense Objectives 2 and 3.[53] This represents a key difference from the unidirectional nature of winning hearts and minds, in which the US military was never forced to confront its own biases or shortcomings when conducting COIN operations.

Finally, the DoD's WPS Plan states clear, concise mission objectives. The policy framework outlines specific Principles, Defense Objectives, and an implementation plan that leverages existing DoD Equities to execute the overarching WPS policy agenda (see Table 4.1 above). This approach enables the DoD to integrate WPS objectives within existing doctrine while also formalizing specific WPS mission sets through standing up the Gender Advisor network and initiating training and education frameworks.

Perhaps the most successful example of DoD WPS implementation so far is the work of the U.S. Indo-Pacific Command operating out of Hawaii, through which the US military has completed operations satisfying all

50 Looney, Joseph D. "Diversity, equity, & inclusion: why does it matter to leadership development?" *The Journal of Character & Leadership Development* 8, no. 2 (2021): 60-67.

51 Heil, Sadia. "Empowered Airmen accelerate change in Air Force training." Air Force News, 7 Sep 2021. Available at https://www.af.mil/News/Commentaries/Display/Article/2766026/empowered-airmen-accelerate-change-in-air-force-training/.

52 U.S. Department of Defense. "Women, Peace, and Security strategic framework and implementation plan." (2020), 8.

53 Indeed, this chapter argues that Defense Objective 1 is critical to distinguishing the WPS policy agenda from winning hearts and minds. Both represent militarized human security, and Defense Objectives 2 and 3 mirror the general engagement approach codified in winning hearts and minds. It is the internal and reflective elements of Defense Objective 1 which render WPS distinct from its predecessor policy, which is why this chapter focuses significant analytical evaluation on this objective.

three of the Defense Objectives.[54] The WPS office within this combatant command conducts internal gender equality training and advises DoD leaders and personnel, in line with Defense Objective 1. With that training foundation established, the office enables partner nation engagement in the region as directed in Defense Objectives 2 and 3. Additional combatant commands are modeling their own WPS staffing and strategic goals after the Indo-Pacific Command example, and the Gender Advisors in Hawaii are proving instrumental to establishing service-level WPS frameworks.[55]

Through deliberately establishing a policy foundation and standing up operational missions based upon that doctrine, the DoD is building a multifaceted foundation from which to execute WPS operations. This differs from the ineffective implementation of a militarized human security approach intended to accomplish broad mission goals outside the military's domain,[56] as was the case with winning hearts and minds. If the DoD can maintain strategic focus and intentional application of the WPS framework, this new phase of militarized human security could function as a force enabler for existing and future military operations.

An Optimistic View? Cautious Recommendations for Future Success

Through a close examination of the policy language initiating two different US militarized human security approaches, this chapter has presented key findings suggesting that implementing the WPS agenda may succeed where winning hearts and minds failed. Rather than exporting US values through force, as under the hearts and minds policy, the WPS agenda enables dialogical engagements that create space for mutual growth and lasting alliances among the US and their partners. For example, the 2022 WPS Congressional Report documents that "U.S. Pacific Air Forces…hosted its first WPS symposium at Joint Base Pearl Harbor-Hickam attended

54 Belsey-Priebe, Maryruth and Tevvi Bullock. "Is the women, peace, and security agenda making progress?" Inkstick Media, 8 Sep 2022. Available at https://inkstickmedia.com/is-the-women-peace-and-security-agenda-making-progress/.

55 Avant, Keiara, Micaela Brancato, and Tahina Montoya. "DAF leads the way on women, peace and security program." *Air Force News*, 30 August 2022. Available at https://www.af.mil/News/Article-Display/Article/3144407/daf-leads-the-way-on-women-peace-and-security-program.

56 Love, Teresa G. "Overseas contingency operations: the danger of fiscal mission creep." *Reporter* 41 (2014): 7.

by 116 virtual and in-person representatives from 20 nations throughout the Indo-Pacific region in March 2021."[57] The WPS agenda will continue accelerating such engagements, which will be critical to maintaining the US' desired status as a "partner of choice"[58] in the future complex strategic environment.

The Defense Objective framework requiring the DoD to exemplify a diverse organization demands an internal reckoning of the institution's own gender inequalities before engaging inequities abroad, thus creating possibilities for transformational policy growth. Specifically, Defense Objective 1 is critical to the long-term success of the WPS agenda: to succeed in any partner engagements where gender equality is a focus, the US must set the example through its own institutional practices. Furthermore, the WPS agenda complements existing DoD mission imperatives and integrates a human security focus within those objectives. This offers an additional possibility for policy success, since integrating WPS within ongoing and future operational campaigns may avoid extreme mission creep compared to the implementation of winning hearts and minds in COIN operations.

Of course, policy frameworks only offer possibilities.[59] People implement policy, and there is always a gap between the intent and the outcome of policy. The WPS framework offers opportunities for success, but DoD leaders must implement the framework judiciously in order to harness these possibilities. This chapter offers three recommendations for achieving success in this next iteration of militarized human security.

1. **The DoD must integrate the WPS agenda holistically across the strategic, operational, and tactical levels of military operations.** DoD leaders and operators must incorporate gender analysis within planning and programming in a coherent, integrated manner alongside all other military planning. If the DoD operationalizes WPS as an ancillary "add women and stir" approach, the entire policy agenda will fail. As a best practice, U.S. DoD leaders should look to Australia's implementation of the

57 The White House. "USG women, peace, and security congressional report." July 2022, 49.
58 U.S. Department of Defense. "Women, Peace, and Security strategic framework and implementation plan." (2020), 10.
59 Atkinson, Kelly E. "Policy and possibilities of humanitarian development: displaced women and peace-building features of the UNHCR." *Refugee Survey Quarterly* 37, no. 4 (2018), 408-439.

WPS agenda and should emulate their admirable strategic plan, the Defence Gender, Peace and Security Mandate 2020-2030.[60]

2. **The DoD must increase funding for Gender Advisor billets and WPS education and training.** At present, existing DoD WPS programming is relatively insignificant compared to the institution's considerable funding resources. As a result, many WPS staff positions are contracted or even volunteer additional duties. This renders the WPS agenda ancillary to DoD priorities and sidelines existing efforts to educate and train the force on the framework. If this era of militarized human security is to succeed, the institution must invest in the framework through legitimately funding assigned personnel positions, education, and training.
3. **The DoD must maintain acute awareness of the considerable critiques surrounding the WPS framework.**[61] In the twenty years since the initiation of UNSCR 1325, scholars and practitioners have acknowledged that gender does just mean women. For example, the original WPS framework equating women with peace essentializes them and reduces their agency. As a result, the policy language emerging from the United Nations now explores Gender, Peace and Security rather than Women, Peace and Security. It is not insignificant that the US has adopted this policy framework utilizing the original, obsolescent language of "women" as opposed to "gender." Again, we must look to Australia as a leader in responsible implementation of this framework.

Australia employs the more appropriate framework of "gender" compared to the narrow categorical employment of the term "women." Furthermore, the Australian Defence Force also incorporates a critical understanding of the role of indigenous populations within all of its strategies and operations, through staffing indigenous liaison officers to acknowledging military use of indigenous land.[62] This approach exemplifies the internal, reflective nature that Defense Objective 1 promotes. The

60 Australian Defence Force. "Defence gender, peace and security mandate: from rhetoric to reality." (2020). Available at https://defence.gov.au/JCG/JSSD/Gender_Peace_Security/20-years-of-wps.asp.

61 Willett, Susan. "Introduction: Security Council Resolution 1325: assessing the impact on women, peace and security." *International peacekeeping* 17, no. 2 (2010): 142-158.

62 Scott, Flight Lieutenant Bel. "Empowered to make a real difference." *Australian Government Defence*, 2 April 2020. Available at https://news.defence.gov.au/people/empowered-make-real-difference.

DoD must be wise enough to follow Australia's example and grasp the opportunity to reflect upon its own institutional stereotypes and biases, despite the attendant cost of acknowledging its own grave errors within the existing organizational structures.[63]

There is always danger in operationalizing women and, more broadly, human security for the purposes of accomplishing the national interest through military operations. Can the US achieve the goals outlined in the WPS policy framework without instrumentalizing women as tools of national power? Moreover, will the DoD learn from previous failures in militarizing human security as the institution engages the WPS agenda? These are questions that require nuanced, critical analysis and thoughtful policy implementation. To implement the WPS agenda responsibly, the DoD must employ skilled academics and practitioners well versed in the policy foundations of this agenda and equip them with funding and resources to integrate the agenda comprehensively. In doing so, the US may learn from past mistakes and succeed in engaging the opportunities available in this newest phase of militarized human security.[64]

References

Adams, Gordon, and Shoon Murray, eds. *Mission creep: the militarization of US foreign policy?*. Georgetown University Press, 2014.

Ashbee, Edward and Steven Hurst. "The Trump foreign policy record and the concept of transformational change." *Global Affairs* 6, no. 1 (2020): 5-19.

Atkinson, Kelly E. "Displacement, gender-based development, and grave violations against children in armed conflict: the case of Iraq." *Journal of Peacebuilding & Development* 17, no. 1 (2022), 108-126.

Atkinson, Kelly E. "Policy and possibilities of humanitarian development: displaced women and peace-building features of the UNHCR." *Refugee Survey Quarterly* 37, no. 4 (2018), 408-439.

Atkinson, Kelly and Alea Nadeem. "Warrior braids and the Air Force women's initiative team – the invisible labor behind diversity, inclusion, and institutional change." *Wild Blue Yonder*, 17 May 2021.

Avant, Keiara, Micaela Brancato, and Tahina Montoya. "DAF leads the way on women, peace and security program." *Air Force News*, 30 August 2022.

63 Mitchell, Ellen. "Reported sexual assaults in US military up 13%." *The Hill*, 1 September 2022. Available at https://thehill.com/policy/defense/3624802-reported-sexual-assaults-in-us-military-up-13/.

64 The views expressed in this chapter are those of the author and do not reflect the official policy or position of the United States Air Force, Department of Defense, or the U.S. Government.

Available at https://www.af.mil/News/Article-Display/Article/3144407/daf-leads-the-way-on-women-peace-and-security-program/.

Australian Defence Force. "Defence gender, peace and security mandate: from rhetoric to reality." (2020). Available at https://defence.gov.au/JCG/JSSD/Gender_Peace_Security/20-years-of-wps.asp.

Bashirov, Galib, and Ihsan Yilmaz. "The rise of transactionalism in international relations: Evidence from Turkey's relations with the European Union." *Australian Journal of International Affairs* 74, no. 2 (2020): 165-184.

Bass, Bernard M., and Ronald E. Riggio. *Transformational leadership*. Psychology press, 2006.

Belsey-Priebe, Maryruth and Tevvi Bullock. "Is the women, peace, and security agenda making progress?" Inkstick Media, 8 Sep 2022. Available at https://inkstickmedia.com/is-the-women-peace-and-security-agenda-making-progress/.

Choi, Seung-Whan, and Idean Salehyan. "No good deed goes unpunished: refugees, humanitarian aid, and terrorism." *Conflict Management and Peace Science* 30, no. 1 (2013): 53-75.

Cohn, Carol, Helen Kinsella, and Sheri Gibbings. "Women, peace and security resolution 1325." *International Feminist Journal of Politics* 6, no. 1 (2004): 130-140.

Dickinson, Elizabeth. "A bright shining slogan: how 'hearts and minds' came to be." *Foreign Policy,* 22 August 2009. Accessible at https://foreignpolicy.com/2009/08/22/a-bright-shining-slogan/.

Dixon, Paul. "'Hearts and minds'? British counter-insurgency from Malaya to Iraq." *Journal of Strategic Studies* 32, no. 3 (2009): 353-381.

Egnell, Robert. "Winning 'hearts and minds'? A critical analysis of counter-insurgency operations in Afghanistan." *Civil Wars* 12, no. 3 (2010): 282-303.

Fishstein, Paul and Andrew Wilder. "Winning hearts and minds? Examining the relationship between aid and security in Afghanistan." *Feinstein International Center,* Tufts University, Medford, MA 50 (2012).

Hazelton, Jacqueline L. "The hearts-and-minds myth: how America gets counterinsurgency wrong." *Foreign Affairs,* 15 July 2021.

Heil, Sadia. "Empowered Airmen accelerate change in Air Force training." Air Force News, 7 Sep 2021. Available at https://www.af.mil/News/Commentaries/Display/Article/2766026/empowered-airmen-accelerate-change-in-air-force-training/.

Hunter, Kyleanne and Emma Moore. "Implementing women, peace, and security guidance to systematically combat gender-based violence in the military." *A Policy Brief from the Athena Leadership Project,* 11 May 2021.

Huntington, Samuel P. *The soldier and the state: The theory and politics of civil–military Relations*. Harvard University Press, 1981.

Kaldor, Mary. *Human security*. Polity, 2007.

Keohane, Jennifer and Kelly Jakes. "Soldiers and scholars: evaluating female engagement teams in the war in Afghanistan." *Women's Studies in Communication* 44, no. 1 (2021): 102-118.

Love, Teresa G. "Overseas contingency operations: the danger of fiscal mission creep." *Reporter* 41 (2014): 7.

Looney, Joseph D. "Diversity, equity, & inclusion: why does it matter to leadership development?" *The Journal of Character & Leadership Development* 8, no. 2 (2021): 60-67.

Manual, Field. "Manual 3-24: Counterinsurgency." *Department of the Army, Washington DC*, 2006.

Mayer, Jane. *The Dark Side: The Inside Story of How the War on Terror Turned into a War on American Ideals*. Anchor, 2009.

Mitchell, Ellen. "Reported sexual assaults in US military up 13%." *The Hill*, 1 September 2022. Available at https://thehill.com/policy/defense/3624802-reported-sexual-assaults-in-us-military-up-13/.

Nell, Karl E. "Hearts-and-minds: a strategy of conciliation, coercion, or commitment?" Army War College, Carlisle Barracks PA (2012).

Petraeus, David H. *Learning counterinsurgency: Observations from soldiering in Iraq*. Army Combined Arms Center, Fort Leavenworth KS (2006).

Petraeus, David H. "Lessons of history and lessons of Vietnam." *The US Army War College Quarterly: Parameters* 16, no. 1 (1986): 27.

Petraeus, David Howell. *The American military and the lessons of Vietnam: a study of military influence and the use of force in the post-Vietnam era*. Princeton University, 1987.

Scott, Flight Lieutenant Bel. "Empowered to make a real difference." *Australian Government Defence*, 2 April 2020. Available at https://news.defence.gov.au/people/empowered-make-real-difference.

The White House. "USG women, peace, and security congressional report." 29 June 2021.

The White House. "USG women, peace, and security congressional report." July 2022.

True, Jacqui. "Explaining the global diffusion of the women, peace and security agenda." *International Political Science Review* 37, no. 3 (2016): 307-323.

United Nations Security Council Resolution 1325, "Women, peace and security." S/RES/1325 (2000). Available at https://documents-dds-ny.un.org/doc/UNDOC/GEN/N00/720/18/PDF/N0072018.pdf?OpenElement.

U.S. Department of Defense. "Women, peace, and security strategic framework and implementation plan" (2020). Available at https://media.defense.gov/2020/Jun/11/2002314428/-1/-1/1/WOMEN_PEACE_SECURITY_STRATEGIC_FRAMEWORK_ IMPLEMENTATION_PLAN.PDF.

United Nations Development Programme. *Human Development Report 1994: New Dimensions of Human Security*, New York (1994).

Vizzard, James W., and Timothy A. Capron. "Exporting General Petraeus's counterinsurgency doctrine: an assessment of the adequacy of Field Manual 3-24 and the US Government's implementation." *Public Administration Review* 70, no. 3 (2010): 485-486.

Von Clausewitz, Carl. "On war." In *On War*. Princeton University Press, 2008.

Willett, Susan. "Introduction: Security Council Resolution 1325: assessing the impact on women, peace and security." *International peacekeeping* 17, no. 2 (2010): 142-158.

5

HOW WOMEN COMBATANTS AFFECT WARTIME RAPE: DISCTINCT EFFECTS IN STATE AND REBEL FORCES

Changwook Ju

Since it was adopted in United Nations Security Council Resolution (UNSCR) 1325, the Women, Peace and Security (WPS) agenda has made the prevention of conflict-related sexual violence against men, women, and children a key part of its imperatives. This issue has been brought to light by survivors, civil society, and dedicated countries, leading to relevant international regulations and a broader concern with the prevention of sexual violence, particularly rape perpetrated during combat. UNSCRs have frequently mentioned women's participation in military operations as vital to prevent wartime rape and safeguard its victims, but concomitantly, such a model of "protective participation" has been criticized as neglecting the "role of gendered power relations in producing the myriad of gendered harms"[1] and thereby hindering women's meaningful participation in peace and security domains.[2]

As such, debates persist as to whether women's participation in armed organizations helps protect non-combatants from wartime rape. Liberal feminist scholars who endorse key elements of critical mass theory, where women's inclusion in and of itself induces broader institutional changes, advocate increasing women's participation in public armies to combat gendered norms and wartime rape perpetration. On the other

1 Heathcote, Gina. 2018. "Security Council Resolution 2242 on Women, Peace and Security: Progressive Gains or Dangerous Development?" *Global Society* 32(4): 374–394.
2 Kirby, Paul, and Laura J. Shepherd. 2016. "The Futures Past of the Women, Peace and Security Agenda." *International Affairs* 92(2): 373–392.

hand, anti-militarist feminists highlight structural, gendered hierarchies and violent militarized masculinity within armed organizations to which women personnel are often pressured and socialized to conform and contribute. Scholars outside of the feminist camp have also documented settings where armed groups with the presence of women combatants[3] do or do not restrain wartime rape and have correspondingly produced inconclusive empirical findings.[4]

This chapter resolves these competing theoretical and empirical predictions by revealing distinct implications of women's combat participation[5] for wartime rape of non-combatants by public and rebel armies: namely, that the presence of women combatants increases the prevalence of wartime rape by state armies but decreases the prevalence of wartime rape by rebel forces. These findings highlight chronic gendered hierarchies in state militaries,[6] while pointing to women's potential for alleviating masculine bonding and misogynistic practices in rebel groups.[7] Furthermore, the findings cast doubt on the pacifying impact of women's

3 Herein, I use the terms *women combatants* and *female combatants* to refer to women serving in combat roles, not merely women in militaries. I recognize that the word *women* is a more inclusive term, whereas the word *female* connotes a biological element. I nonetheless use these two words interchangeably as some scholars employ terms like "female combatancy" and "female-intensive insurgencies" in the relevant literature. See Loken, Meredith. 2017. "Rethinking Rape: The Role of Women in Wartime Violence." *Security Studies* 26(1): 60–92; and Wood, Elisabeth Jean. 2006. "Variation in Sexual Violence During War." *Politics & Society* 34(3): 307–342, 330.

4 Cohen, Dara Kay, Amelia Hoover Green, and Elisabeth Jean Wood. 2013. *Wartime Sexual Violence: Misconceptions, Implications, and Ways Forward*. Washington, DC: United States Institute of Peace; Hoover Green, Amelia. 2017. "Armed Group Institutions and Combatant Socialization: Evidence from El Salvador." *Journal of Peace Research* 54(5): 687–700; Hoover Green, Amelia. 2018. *The Commander's Dilemma: Violence and Restraint in Wartime*. Ithaca, NY: Cornell University Press; Loken, "Rethinking Rape"; Mehrl, Marius. 2022. "Female Combatants and Wartime Rape: Reconsidering the Role of Women in Armed Conflict." *Armed Forces & Society* 48(2): 464–479; Wood, "Variation in Sexual Violence During War"; Wood, Elisabeth Jean. 2009. "Armed Groups and Sexual Violence: When Is Wartime Rape Rare?" *Politics & Society* 37(1): 131–161.

5 By women's combat participation, I mean their participation in armed organizations as combatants. Combatants are conceptualized herein as fighters who undergo combat training and are deployed to front-line battlegrounds. Women's combat roles differ from their non-combat roles, which include intelligence or spies in clandestine operations, as well as various support, transportation, and logistical roles.

6 Enloe, Cynthia H. 2000. *Maneuvers: The International Politics of Militarizing Women's Lives*. Berkeley, CA: University of California Press; Goldstein, Joshua S. 2001. *War and Gender: How Gender Shapes the War System and Vice Versa*. Cambridge: Cambridge University Press; Sjoberg, Laura, and Sandra Via. 2010. *Gender, War, and Militarism Feminist Perspectives*. Santa Barbara, CA: Praeger.

7 Hoover Green, "Armed Group Institutions and Combatant Socialization"; Hoover Green, *The Commander's Dilemma*; Wood, "Variation in Sexual Violence During War"; Wood, "Armed Groups and Sexual Violence."

military participation in preventing state-directed rape, while emphasizing a gender-sensitive perspective and women's substantive engagement in military operations, which are better reflected in rebel groups.

This chapter first surveys relevant theoretical arguments and, using evidence from novel empirical analyses, draws the implications of women's combat involvement for the WPS agenda and militaries' approach toward wartime rape. It reaches these conclusions by revisiting two extant empirical studies on women's military participation and wartime rape by belligerents (both regular armies and rebel groups). Methodologically, it deploys an estimation technique that addresses the disproportionate number of "zero" observations in wartime rape data, which conflate the true absence of rape (that is, rape did not occur) with its unknown presence (that is, rape was unreported / undocumented despite its empirical existence). To address this ambiguity of such "dual-nature" zeros, I differentiate the two types of zero observations on a probability scale and then estimate the prevalence of wartime rape conditional on the statistical partitioning of its two zero types.

Women Combatants and Wartime Rape

Theoretical Predictions

There has been no shortage of discourse surrounding women's contributions and experiences in war. Researchers have vigorously studied women's participation in armed organizations at a cross-national level and have discovered interesting patterns of women's involvement in warfare and their wartime experiences.[8] However, women's potential for pacifying wartime human rights violations, especially rape, remains debated.

8 Haer, Roos, and Tobias Böhmelt. 2018. "Girl Soldiering in Rebel Groups, 1989–2013: Introducing a New Dataset." *Journal of Peace Research* 55(3): 395–403; Henshaw, Alexis Leanna. 2016. "Where Women Rebel: Patterns of Women's Participation in Armed Rebel Groups 1990–2008." *International Feminist Journal of Politics* 18(1): 39–60; Henshaw, Alexis *et al.* 2019. "Understanding Women at War: A Mixed-Methods Exploration of Leadership in Non-State Armed Groups." *Small Wars & Insurgencies* 30(6–7): 1089–1116; Thomas, Jakana L., and Kanisha D. Bond. 2015. "Women's Participation in Violent Political Organizations." *American Political Science Review* 109(3): 488–506; Thomas, Jakana L, and Reed M Wood. 2018. "The Social Origins of Female Combatants." *Conflict Management and Peace Science* 35(3): 215–232; Wood, Reed M, and Jakana L Thomas. 2017. "Women on the Frontline: Rebel Group Ideology and Women's Participation in Violent Rebellion." *Journal of Peace Research* 54(1): 31–46.

Feminist analyses have focused on public armies, or explicitly on Western state armies,[9] to proffer different predictions regarding the implications of women's combat participation for wartime rape. On the one hand, liberal feminist scholars advocate the inclusion of women in public armies to shift gendered norms and violent practices inside and outside their ranks and to decrease rape and other types of violence during wartime.[10] These ideas are predicated on the premise in critical mass theory[11] that the greater inclusion of women would unambiguously decrease the persistence of wartime rape by inducing positive institutional and social-cultural improvements in armed forces.[12] If women comprise militaries "in large numbers, at every rank, in every department," Card writes, "rape is an unlikely instrument of heterosexual peer bonding."[13] Liberal feminists rely crucially on this "essentialist" view of women's peaceful nature in addressing the problem of wartime rape.[14]

On the other hand, anti-militarist feminists promulgate contrary accounts based on structural sexism, systematic exclusion, and the social construction of gendered hierarchies in militaries. They suggest that the integration of women into militaries tends to *reinforce* the gendered order and heterosexual, militarized masculinity that elicit rape and sexual abuse

---

9 Duncanson, Claire, and Rachel Woodward. 2016. "Regendering the Military: Theorizing Women's Military Participation." *Security Dialogue* 47(1): 3–21.

10 Card, Claudia. 1996. "Rape as a Weapon of War." *Hypatia* 11(4): 5–18; Duncanson and Woodward, "Regendering the Military," 5–6; MacKinnon, Catharine A. 2006. *Are Women Human? And Other International Dialogues*. Cambridge, MA: Belknap Press of Harvard University Press; Stiehm, Judith Hicks, ed. 1996. *It's Our Military, Too! Women and the U.S. Military*. Philadelphia, PA: Temple University Press.

11 For broader discussions on critical mass theory in miliary and political domains, see Beckwith, Karen. 2007. "Numbers and Newness: The Descriptive and Substantive Representation of Women." *Canadian Journal of Political Science* 40(1): 27–49; and Bystydzienski, Jill M. 1993. "Women in Groups and Organizations: Implications for the Use of Force." In *Women and the Use of Military Force*, eds. Ruth H. Howes and Michael R. Stevenson. Boulder, CO: Lynne Rienner Publishers, 39–52.

12 For example, Annica Kronsell contends that "the inclusion of even a small percentage of women makes all the difference in the world because it makes gender and masculine norms visible. Even when women remain a minority and are perceived as tokens within an institution, the presence of women calls attention to the norms of the dominant culture because women are understood to represent 'difference', i.e. the dominant majority become aware of themselves and what binds them together in terms of explicitly gendered norms." See Kronsell, Annica. 2005. "Gendered Practices in Institutions of Hegemonic Masculinity." *International Feminist Journal of Politics* 7(2): 280–298, 286.

13 Card, "Rape as a Weapon of War," 12.

14 Mobekk, Erin. 2010. "Gender, Women and Security Sector Reform." *International Peacekeeping* 17(2): 278–291; Kirby and Shepherd, "The Futures Past of the Women, Peace and Security Agenda," 375.

in wartime,[15] including those directed by women fighters who are often forced to partake in or socialized to have agency in collective rape alongside their male comrades-in-arms.[16] These ideas emphasize that armed forces are inherently grounded on hyper-masculine, sexist foundations that resist changes and reforms in the direction of what liberal feminists believe. Rather, from anti-militarist feminists' vantage point, the presence of women in armies would not decrease rape in wartime and may even increase it.[17]

In short, there is a conspicuous divide among feminists as to whether women's military engagement signals potential structural advances with gender equality and empowerment that moderate rape during war, or, conversely, co-optation and subjugation that exacerbate the problem of wartime rape.[18] Further complicating these diverging theoretical predictions, scholars outside the feminist camp have identified contexts

15 Enloe, *Maneuvers*; Goldstein, *War and Gender*; Reardon, Betty. 1985. *Sexism and the War System*. New York: Teachers College Press; Shepherd, Laura J. 2007. "'Victims, Perpetrators and Actors' Revisited: Exploring the Potential for a Feminist Reconceptualisation of (International) Security and (Gender) Violence." *British Journal of Politics and International Relations* 9(2): 239–256; Sjoberg and Via, *Gender, War, and Militarism Feminist Perspectives*.

16 Sjoberg, Laura. 2007. "Agency, Militarized Femininity and Enemy Others: Observations from the War in Iraq." *International Feminist Journal of Politics* 9(1): 82–101; Sjoberg, Laura. 2016. *Women as Wartime Rapists: Beyond Sensation and Stereotyping*. New York: New York University Press.

17 Duncanson and Woodward, "Regendering the Military," 6.

18 Although outside the purview of this chapter, similar theoretical and empirical debates exist in the post-conflict peacekeeping literature, implying the importance of the link between female personnel and the conflict-related indignities of those outside military ranks. Some scholars argue for increasing the presence of women peacekeeping personnel to combat sex crimes. See, for example, Bridges, Donna, and Debbie Horsfall. 2009. "Increasing Operational Effectiveness in UN Peacekeeping: Toward a Gender-Balanced Force." *Armed Forces & Society* 36(1): 120–130; Defeis, Elizabeth F. 2008. "U.N. Peacekeepers and Sexual Abuse and Exploitation: An End to Impunity." Washington University *Global Studies Law Review* 7(2): 185–214, 199; and DeGroot, Gerard J. 2001. "A Few Good Women: Gender Stereotypes, the Military and Peacekeeping." *International Peacekeeping* 8(2): 23–38. Others highlight the caveats and compelling reasons for its bounded effects. See, for example, Karim, Sabrina, and Kyle Beardsley. 2016. "Explaining Sexual Exploitation and Abuse in Peacekeeping Missions: The Role of Female Peacekeepers and Gender Equality in Contributing Countries." *Journal of Peace Research* 53(1): 100–115, 103–105; Kirby and Shepherd, "The Futures Past of the Women, Peace, and Security Agenda"; Mobekk, Eirin. 2010. "Gender, Women and Security Sector Reform." *International Peacekeeping* 17(2): 278–291, 285–287; and Valenius, Johanna. 2007. "A Few Kind Women: Gender Essentialism and Nordic Peacekeeping Operations." *International Peacekeeping* 14(4): 510–523. Sandra Whitworth especially notes that as militaries often rely on militarized masculinity to mobilize and motivate soldiers, there is a resistance to the inclusion of female characters, and women personnel often repress their feminine characteristics, such as empathy for victims of rape and tendencies to stop it. See Whitworth, Sandra. 2004. *Men, Militarism, and UN Peacekeeping: A Gendered Analysis*. Boulder, CO: Lynne Rienner. Olivera Simić points to the mixed results on women's pacifying effect on sexual violence in peacekeeping operations. See Simić, Olivera. 2010. "Does the Presence of Women Really Matter? Towards Combating Male Sexual Violence in Peacekeeping Operations." *International Peacekeeping* 17(2): 188–199.

where the presence of female combatants in armed organizations does or does not reduce wartime rape. Both Hoover Green and Wood explain the restraint of wartime rape by some rebel groups with substantial proportions of women combatants,[19] which implies women's potential for undermining masculine bonding and misogynistic training in armed organizations. However, Cohen documents women combatants who actively engaged in gang rape alongside their male counterparts, suggesting that "norms of gender equality should not be assumed from the mere presence of women in armed groups."[20] Thus, women appear to help some rebel organizations reduce wartime rape, while in other cases, it seems to require far more extensive and substantive changes.

Empirical Works

Empirical investigations have also produced differing findings on the relationship between women's combat participation and wartime rape of non-combatants by belligerent groups. To begin, Loken uses "female combatancy" to indicate women's involvement in combat roles[21] and finds that it has no effect on state- and rebel-perpetrated rape during war.[22] Consequently, she contends that women's integration generally does not subdue gendered relations and misogynistic culture in armed organizations to which women fighters are pressured to conform.

More recent studies paint a more complicated picture. For example, in a reanalysis of Loken's study focusing specifically on rebel groups, Mehrl adjusts for two main modeling issues in Loken's estimations and finds an evident inverse relationship between female combatancy and wartime rape by rebel groups.[23] In his extension analysis, however, Mehrl expects but fails to find an effect such that female combatancy lessens wartime rape by rebel groups with less misogynistic norms, which was measured by whether

19 Hoover Green, "Armed Group Institutions and Combatant Socialization"; Hoover Green, *The Commander's Dilemma*; Wood, "Variation in Sexual Violence During War"; Wood, "Armed Groups and Sexual Violence."

20 Cohen, Dara Kay. 2013. "Female Combatants and the Perpetration of Violence: Wartime Rape in the Sierra Leone Civil War." *World Politics* 65(3): 383–415, 408.

21 Loken indicates that she uses "conservatively coded" women's combat participation because women "may move in and out of a myriad of positions including support, transport, and logistics, and they may even return to civilian life during war." See Loken, "Rethinking Rape," 72.

22 Loken, "Rethinking Rape."

23 Mehrl, "Female Combatants and Wartime Rape."

or not women held commanding positions.[24] In short, the literature has yielded mixed and unexpected results on women's combat participation and wartime rape.

This chapter revisits these two empirical studies to adjudicate the inconclusive theoretical and empirical predictions on the relationship between women's combat participation and wartime rape. Therefore, below, it lays out the foundation for its research framework by describing the main variables of interest and estimation techniques used in the studies, as well as corresponding methodologies.

Revisiting Women Combatants and Wartime Rape

Before proceeding to empirical examinations, I explicate existing empirical studies and then motivate my methods.

Two Existing Empirical Studies

Loken's cross-national study examines the impact of female combatancy on wartime rape using an original dataset on women's involvement as fighters in state and rebel forces in 86 major civil conflicts from 1980 to 2009, including 983 conflict–year observations.[25] She provides four female combatancy variables: (i) "female combatancy in rebel groups"; (ii) "female combatancy in state forces"; (iii) "female combatancy in rebel groups including girl soldiers and/or mixed roles"; and (iv) "female combatancy in state forces including girl soldiers and/or mixed roles." [26] These variables have three levels: none/low (0); moderate (1); and high (2). The first and second variables are used in Loken's main statistical models, while the third and fourth variables "account for cases in which women's roles are mixed or indistinguishable from other supportive duties," such as in the case of Israeli female conscripts who constitute only about 3 per cent of combat forces but arguably serve as broader symbols in Israel's military affairs.[27]

Loken uses the data on rape by state and non-state forces during civil war that Cohen collected based on the United States Department of

24 Ibid., 473.
25 Loken, "Rethinking Rape."
26 Ibid., 70–73.
27 Loken, "Rethinking Rape," 72.

State (US DoS) Human Rights Country reports.[28] Cohen codes wartime rape by belligerents using four-part ordinal measures, which range from 0 when there was "no mention" of wartime rape in the reports, to 3 when rape was reported as "massive" or "systematic."[29] Loken employs two four-level modified ordinal variables for state- and rebel-perpetrated wartime rape: (1) "the highest level of rape insurgent groups perpetrated in each year" and (2) "the highest level of rape state actors perpetrated in each year."[30] She employs ordered probit (OP) regressions with robust standard errors clustered by conflict[31]; OP models fit the ordinal variables by keeping the ordering of their coding (i.e., 0; 1; 2; 3) while making no assumptions about the substantive distances between.

Mehrl re-examines Loken's study on women combatants and wartime rape, focusing exclusively on rebel groups.[32] He uses the same female combatancy variable for rebel groups and OP regression models used in Loken's research. Based on his findings, Mehrl argues that the presence of female rebel combatants reduces wartime rape. He also extends Loken's study. He analyzes the impact of female rebel participation on wartime rape, conditional on command strength and misogynistic norms, with the latter proxied by a binary variable that indicates whether rebel groups include *Female Commanders* in their leadership.[33] Later, I engage with only the interaction model about misogynistic group norms, which is the major emphasis in the works of both Loken and Mehrl.[34] For this analysis, Mehrl "binarize[s]" the female combatancy variable by combining the moderate (1) and high (2) categories into 1.[35]

Importantly, Mehrl addresses two major issues with Loken's statistical models. First, Loken's main independent variable and three control variables are treated as continuous in her estimation models despite all of them being

28 Cohen, Dara Kay. 2016. *Rape during Civil War*. Ithaca, NY: Cornell University Press. Loken considered but abstained from relying on the Sexual Violence in Armed Conflict (SVAC) dataset from Cohen and Nordås's 2014 publication, "Sexual Violence in Armed Conflict: Introducing the SVAC Dataset, 1989–2009," because she is specifically interested in the prevalence of one type of sexual violence, i.e., rape, and because of dim prospects for matching the SVAC data with the limited data on women's participation as combatants that was available at the time. See Loken, "Rethinking Rape," 75.

29 Cohen, *Rape during Civil War*, 66.

30 Loken, "Rethinking Rape," 75.

31 For descriptions of control variables, see Loken "Rethinking Rape," 74–75.

32 Mehrl, "Female Combatants and Wartime Rape."

33 Henshaw *et al.*, "Understanding Women at War."

34 Loken, "Rethinking Rape," 80–90; Mehrl, "Female Combatants and Wartime Rape," 473–475.

35 Mehrl, "Female Combatants and Wartime Rape," 473.

categorical. Specifically, her models treat *Female Rebels* (None; Moderate; High), *State Failure* (None; Regime Change; Limited; Substantial; Complete), *Ethnic War* (No; Ambiguous; Yes), and *Conflict Aim* (Government; Ambiguous; Secession) in a manner that assumes "a linear relationship between categories where there is none."[36] However, a shift from, for example, the 'None' category to the 'Moderate' category in *Female Rebels* may not be substantially identical or proportional to a change from the 'Moderate' category to the 'High' category, nor do these two changes in the variable have the same effect on rape. Moreover, *Ethnic War* and *Conflict Aim* do not feature clear rank ordering. Second, Loken's analysis incorporates a continuous variable indicating each observation's calendar year. This modeling practice imposes a strong assumption on the data, namely that the baseline hazard of wartime rape monotonically grows, declines, or remains constant over time. This assumption, however, is not supported by her data; although observations in earlier years contain less reported information about rape (that is, the 0 level), there is no evident monotonic trend in the non-zero levels of wartime rape over time (see Figure 5.1).

Mehrl addresses these two concerns while otherwise adhering to Loken's selection of data, sample, variables, and estimation technique. First, he separates each of the four variables into factor dummies. Second, he employs more flexible modeling techniques to account for non-linear temporal trends rather than incorporating years in a linear fashion like in previous analyses. He primarily uses year splines, which enable the recorded prevalence of wartime rape to fluctuate nonlinearly with time.

Methods

Both Loken and Mehrl model the ordinal measures of wartime rape using conventional OP regression models.[37] However, as plotted in Figure 5.1, the ordinal data contain enormous "zero" observations coded by Cohen when there was "no mention" of wartime rape in the US DoS Human Rights Country reports.[38] While Cohen clarifies that the coding of 0 does not necessarily imply no rape,[39] it conflates a lack of rape and no mention of rape under its zero-level prevalence, inflating zero observations without

36 Ibid., 470.

37 Loken, "Rethinking Rape"; Mehrl, "Female Combatants and Wartime Rape."

38 Cohen, *Rape during Civil War*, 66.

39 Woodbridge, Jeffrey M. 2013. *Introductory Econometrics: A Modern Approach*. 5th edition. Mason, OH: Cengage Learning, 228.

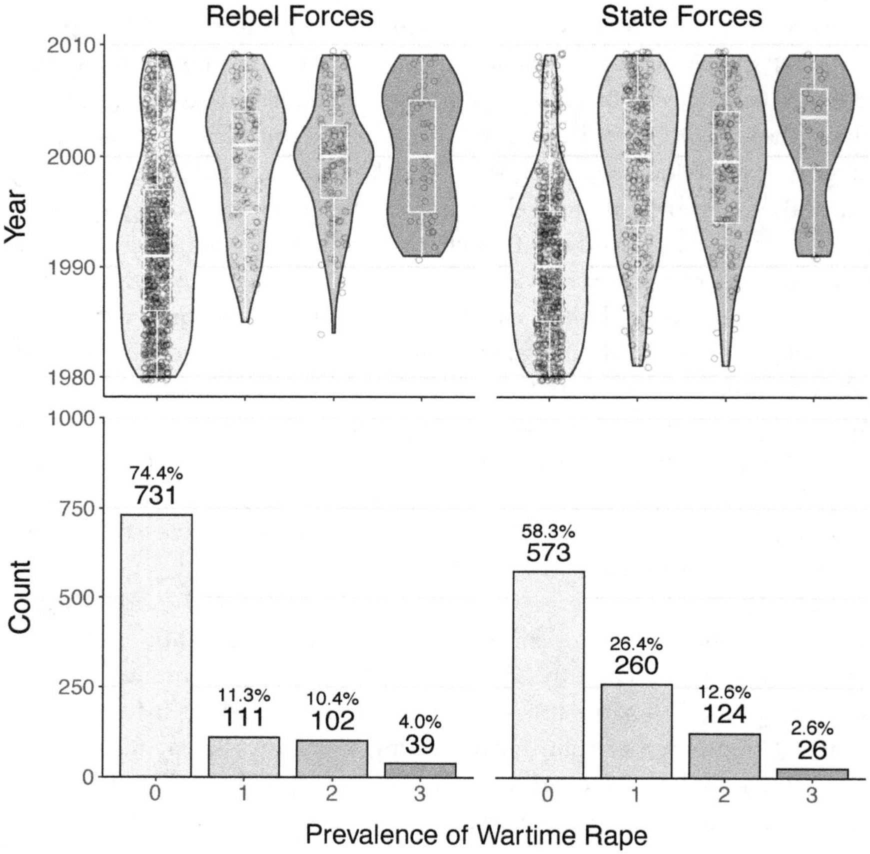

Figure 5.1. Distribution of Ordinal Data on Wartime Rape
Note: Prevalence scale of 0 = No mention of rape in US DoS human rights reports ("None"); 1 = "Isolated" reports of rape; 2 = Reports of widespread rape ("Numerous"); 3 = Reports of systematic rape ("Massive"). Data points are "jittered" to avoid overlaps.

clearly sourced indications of little to no wartime rape.[40] As Krüger and Nordås write, "Zero-level observations are challenging in that they remain unclear with regard to the true level of abuse: was there really no practice of sexual violence, or did actual abuses go unreported?"[41]

40 As a rule of thumb, values for which scholars cannot locate sources of zero-level rape indication should be classified as NAs or under an 'Unknown' category rather than 0s because no mention of rape in particular human rights reports does not imply a true rape level of 0.
41 Krüger, Jule, and Ragnhild Nordås. 2020. "A Latent Variable Approach to Measuring Wartime Sexual Violence." *Journal of Peace Research* 57(6): 728–739, 737.

Zeros in wartime rape data, or the lowest ordinal scale of prevalence, are of two classes: i) rape did not occur (hence no indications), or ii) rape occurred but was undocumented in the US DoS Human Rights Country reports due to its relatively small magnitude and/or various reporting issues (and reported rape may even go unmentioned or undocumented).[42] To address these two sources of zeros, I divide the 0 category further and herein call them "true 0s" (a true absence of rape) and "inflated 0s" (an unknown presence of rape). Because "true 0s" mix with "inflated 0s," zero observations comprise a disproportionately large fraction of wartime rape data, while one is hard-pressed to tell their genuine nature. Traditional OP models, wherein zeros are modeled as solely indicating a lack of rape, cannot model the preponderance of zeros, nor can they distinguish between the two sources of zeros, incurring considerable information loss and inferential bias.[43]

To rectify this, arguably the most dominant yet intractable issue with ordinal wartime rape data and corresponding estimation, I employ zero-inflated ordered probit (ZIOP) regressions, which are referred to as "mixture" models that regard zeros as arising from two different processes.[44] In other words, ZIOP estimations statistically differentiate the two classes of zeros, a true absence of rape (that is, non-occurrence) and an unknown presence of rape (that is, no reporting/documentation despite empirical existence), in two underlying processes, as illustrated in Figure 5.2. First, the inflation stage uses one set of covariates to split the two classes of 0s and predict the probability of inflated 0s, that is, the unknown

42 Measuring and estimating sexual violence in war (including rape) suffers from significant reporting and documentation challenges. See, inter alia, Cohen and Nordås, "Sexual Violence in Armed Conflict"; Cohen, Dara Kay, and Amelia Hoover Green. 2012. "Dueling Incentives: Sexual Violence in Liberia and the Politics of Human Rights Advocacy." *Journal of Peace Research* 49(3): 445–458; Krüger and Nordås, "A Latent Variable Approach to Measuring Wartime Sexual Violence"; Leiby, Michele. 2009. "Digging in the Archives: The Promise and Perils of Primary Documents." *Politics & Society* 37(1): 75–99; Leiby, Michele L. 2012. "The Promise and Peril of Primary Documents: Documenting Wartime Sexual Violence in El Salvador and Peru." In *Understanding and Proving International Sex Crimes*, eds. Morten Bergsmo, Alf B. Skre, and Elisabeth Jean Wood. Beijing: Torkel Opsahl Academic EPublisher, 315–366; Roth, Françoise, Tamy Guberek, and Amelia Hoover Green. 2011. *Using Quantitative Data to Assess Conflict-Related Sexual Violence in Colombia: Challenges and Opportunities*. Benetech Human Rights Program and Corporación Punto de Vista; and Wood "Variation in Sexual Violence During War."

43 Bagozzi, Benjamin E., Daniel W. Hill, Will H. Moore, and Bumba Mukherjee. 2015. "Modeling Two Types of Peace: The Zero-Inflated Ordered Probit (ZiOP) Model in Conflict Research." *Journal of Conflict Resolution* 59(4): 728–752; Harris, Mark N., and Xueyan Zhao. 2007. "A Zero-Inflated Ordered Probit Model, with an Application to Modelling Tobacco Consumption." *Journal of Econometrics* 141(2): 1073–1099.

44 Bagozzi *et al.*, "Modeling Two Types of Peace," 733.

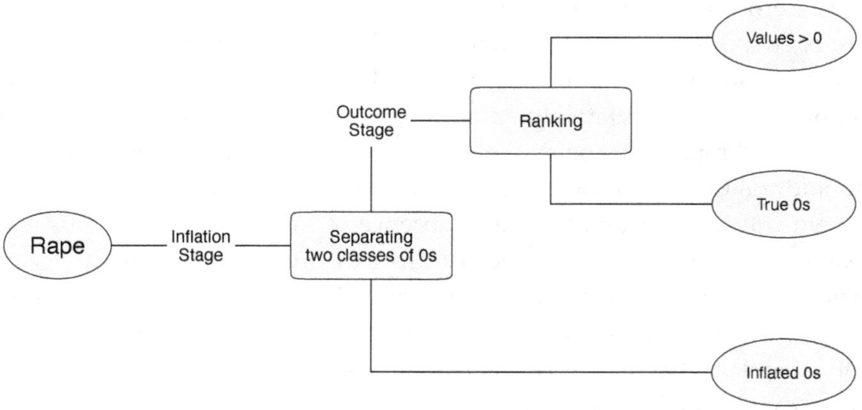

Figure 5.2. The Underlying Processes of ZIOP Models

presence of wartime rape. Second, the outcome stage uses another set of covariates to remove inflated 0s and predict the prevalence of wartime rape incorporating true 0s.

Simply put, ZIOP estimations differentiate the two types of zeros on a probability scale (inflation stage) and then estimate the level of wartime rape using "true 0s" and values larger than 0, thereby generating two sets of estimates from these two latent stages. While the zeros in question are not random and are likely correlated with an ensemble of cross-sectional and temporal factors that might already induce bias in statistical inference, ZIOP estimations account for such "observable and latent factors" that underlie disproportionate zero observations and the probability of their two distinct types.[45] Consequently, ZIOP estimators impart "substantively rich information about the two types of zero observations" through their inflation-stage estimates.[46]

In terms of interpretation, the inflation model of ZIOP regressions provides information regarding the chance of the potentially unknown presence of rape; a positive coefficient indicates a higher likelihood that an observation involves rape that goes unreported (that is, "inflated 0"), and correspondingly, a lower likelihood that an observation actually entails the non-occurrence of rape (that is, "true 0"). The outcome model of ZIOP regressions has an equivalent substantive meaning as typical OP regressions, save for a conditional statement: a positive coefficient suggests an increase in rape levels, conditional on rape being able to potentially occur

45 Bagozzi *et al.*, "Modeling Two Types of Peace," 729–730.
46 Ibid., 730.

and be reported. In other words, a positive coefficient means an increased likelihood of observing rape in the first place and then an increase in the prevalence level of rape across true 0s and non-zero categories.

Because no mention of wartime rape (that is, Not Available, or NA) is treated as its zero-level prevalence, I also conduct what I call "NA analysis," in which I replace all 0s with NAs and then fit the data with OP models used in original analysis. To be sure, NA analysis merits two caveats: First, because of zero abundance, sample sizes fall dramatically (see Table 5.1); hence, results should be interpreted with some caution. Second, there may be "true 0s"; therefore substituting 0s for NAs "likely discards relevant conflict observations and produces selection bias."[47] I thus justify NA analysis as an *auxiliary* comparison tool, presenting its estimates alongside original and zero-inflated ones; the rationale for NA analysis is to determine whether original OP models without potentially inflated 0s predict comparable results as ZIOP outcome models.

Unlike NA analysis, which necessarily loses observations models retain all observations used in original models (see Table 5.1). In principle, I estimate ZIOP models with all original covariates appearing in both the outcome and inflation stages. While certain variables in the inflation stage are omitted in some of the ZIOP models in the process of addressing convergence issues, all variables remain intact in the outcome stage, which resembles the original OP models.

Table 5.1 summarizes the statistical models analyzed herein and reports the Akaike Information Criterion (AIC) and bias-corrected AIC (AICc) scores to formally compare model fit between the OP and ZIOP models estimated for my analysis.[48] All ZIOP models show much lower AIC(c) statistics than original models do; ZIOP models fit wartime rape data far better than conventional OP models. Note that I do not consider using the (bias-corrected) Vuong closeness test[49] to compare model fitting because it has been shown to be inappropriate for testing the model fit of zero-inflated estimations.[50]

47 Bagozzi, Benjamin E. 2015. "Forecasting Civil Conflict with Zero-Inflated Count Models." *Civil Wars* 17(1): 1–24, 7.
48 The AIC(c) statistics of original and zero-inflated models cannot be compared with those of NA models, which have reduced sample sizes.
49 Desmarais, Bruce A., and Jeffrey J. Harden. 2013. "Testing for Zero Inflation in Count Models: Bias Correction for the Vuong Test." *Stata Journal* 13(4): 810–835; Vuong, Quang H. 1989. "Likelihood Ratio Tests for Model Selection and Non-Nested Hypotheses." *Econometrica* 57(2): 307–333.
50 Wilson, Paul. 2015. "The Misuse of the Vuong Test for Non-Nested Models to Test for Zero-Inflation." *Economics Letters* 127(C): 51–53.

				OP		NA	ZIOP	
Study	Table/ Model	Dependent Variable	Independent Variable	n	AIC (AICc)	n	n	AIC (AICc)
Loken (2017)	Table 3 Model 2	Wartime Rape by Rebel Forces	Female Combatancy	945	1,245 (1,246)	239	945	1,159 (1,161)
Loken (2017)	Table 3 Model 3	Wartime Rape by State Forces	Female Combatancy	692	1,046 (1,047)	220	692	1,019 (1,022)
Mehrl (2022)	Table 1 Model 6	Wartime Rape by Rebel Forces	Female Combatancy: Moderate	867	1,038 (1,039)	210	867	961 (965)
Mehrl (2022)	Table 1 Model 6	Wartime Rape by Rebel Forces	Female Combatancy: High	867	1,038 (1,039)	210	867	961 (965)
Mehrl (2022)	Model A13	Wartime Rape by Rebel Forces	Female Combatancy × Female Commanders	747	910 (912)	188	747	842 (847)

Table 5.1. Models and AIC(c) Statistics
Note: "n" denotes the number of observations estimated in statistical models. AIC(c) statistics are rounded to the nearest integer. A *lower* value of AIC(c) indicates evidence for a *better* model fit.

Results and Discussion

Figure 5.3 reports the statistical results of original, NA, and zero-inflated analyses on female combatancy and rape of non-combatants during civil war. In presenting my findings, I probe existing theoretical assumptions and point to explanations that do or do not remain robust, alongside the reasons why they do or do not. In the following analyses, all coefficients are estimated using robust standard errors clustered by conflict, sample sizes vary depending on the study and model (see Table 5.1). In addition, all the original covariates in Loken's and Mehrl's studies are included in all estimated models.[51]

First, contrary to Loken's null finding on the relationship between female combatancy and wartime rape *in toto*,[52] I find that the presence of women combatants increases the prevalence of wartime rape by state forces. My NA model and ZIOP outcome models for Loken's study demonstrate that women's combat participation in state forces leads to the groups' higher level of wartime rape. My findings indicate that liberal feminists' ideas based on the logic of critical mass theory do not hold true in state militaries, as women combatants in state armies remain rare in absolute terms and in comparison to numbers in non-state armed forces.[53] While women's influence is limited in male-dominated state armies, my findings lend credence to the anti-militarist feminists' contention that the

51 Loken, "Rethinking Rape"; Mehrl, "Female Combatants and Wartime Rape."
52 Loken, "Rethinking Rape."
53 Goldstein, *War and Gender*; King, Anthony. 2016. "The Female Combat Soldier." *European Journal of International Relations* 22(1): 122–143.

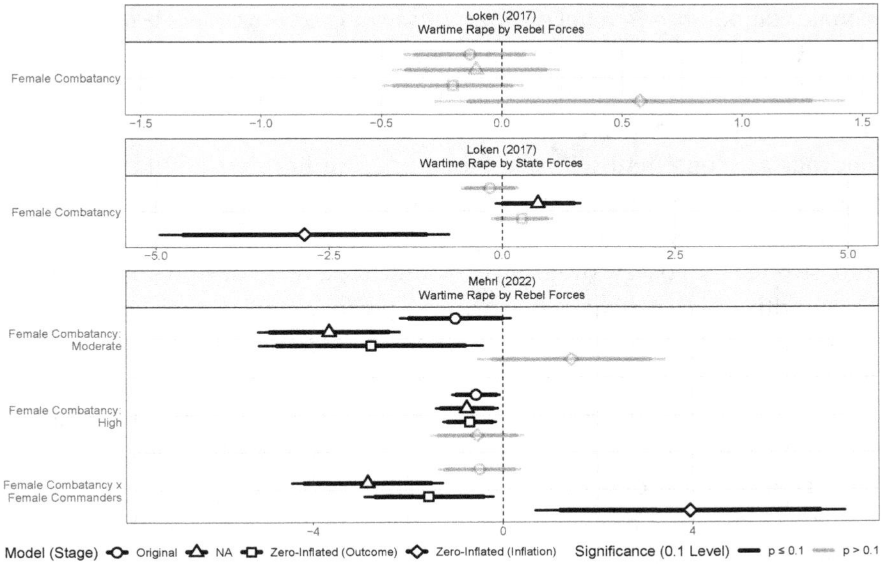

Figure 5.3. Women Combatants and Wartime Rape
Note: Thick and thin lines indicate 90 per cent and 95 per cent confidence intervals around corresponding raw coefficient estimates, respectively.

inclusion of women combatants tends to buttress gendered hierarchies and heterosexual, militarized masculinity that elicit (collective) rape by both male and female combatants during war.[54]

Second, contra the results concerning state armies, I find that the presence of women combatants decreases the prevalence of wartime rape by rebel forces. On the one hand, female combatancy, as a standalone variable, appears to allay wartime rape by rebel forces. Even if my estimation models for Loken's study fall a little short of statistical significance, the effect size in the outcome model grows larger than the original model in the negative direction. In effect, my NA and ZIOP outcome models for Mehrl's study manifest that moderate and high levels of female combatancy are negatively associated with wartime rape perpetrated by rebel forces. On the other hand, this pacifying impact is conditioned by less misogynistic norms in rebel groups. My NA and ZIOP outcome models for Mehrl's study bring confidence to his intuitive yet unsubstantiated theoretical expectation that

54 See, among many others, Enloe, *Maneuvers*; Goldstein, *War and Gender*; Reardon, *Sexism and the War System*; Shepherd, "Victims, Perpetrators and Actors' Revisited"; Sjoberg, "Agency, Militarized Femininity and Enemy Others"; Sjoberg, *Women as Wartime Rapists*; Sjoberg and Via, *Gender, War, and Militarism Feminist Perspectives*.

female combatancy in rebel groups with female commanders is particularly effective at reducing the prevalence of rebel-perpetrated wartime rape.

Strong theoretical and historical foundations undergird the negative impact of women combatants on wartime rape by rebel groups. My findings are consistent with the theoretical implications distilled from the works of Hoover Green and Wood, who provide significant evidence that "female-intensive insurgencies" with gender-equal norms and practices and stronger ideology programs restrain rape against non-combatants.[55] Even without these mechanisms, the presence of women combatants in rebel groups is said to have an independent impact on wartime rape through sociocultural processes, as in the case of the Lord's Resistance Army in northern Uganda, whose strategic forced recruitment of women and girls and their resulting marriages led to intra-rebel socialization of both genders and restraint in rape outside the group's ranks.[56] Similarly, there is qualitative evidence that women's integration into the People's Liberation Army in Nepal shifted internal gendered norms and reduced the group's violence against women.[57] These narratives and my findings underscore the ability of rebel groups to re-socialize combatants away from malignant gendered hierarchies and misogynistic practices,[58] resulting in considerably less rape during civil war.

Interestingly, my NA analysis corroborates all these findings as well as the so-called "non-findings" from ZIOP outcome models (see Figure 5.3), bolstering credence to my claim that ungrounded excessive 0s matter. That is, traditional estimation methods perform well absent potentially inflated 0s.

Finally, my ZIOP inflation estimates suggest that female combatancy decreases the likelihood of unreported wartime rape by state forces (that is, inflated 0s), and that female combatancy and leadership increase the likelihood of unreported wartime rape by rebel forces. How might we

55 Hoover Green, "Armed Group Institutions and Combatant Socialization"; Hoover Green, *The Commander's Dilemma*; Wood, "Variation in Sexual Violence During War," 330; Wood, "Armed Groups of Sexual Violence"; Wood and Thomas, "Women on the Frontline."

56 Kramer, Sophie. 2012. "Forced Marriage and the Absence of Gang Rape: Explaining Sexual Violence by the Lord's Resistance Army in Northern Uganda." *Journal of Politics and Society* 23(1): 11–49; Carlson, Kristopher, and Dyan Mazurana. 2008. *Forced Marriage within the Lord's Resistance Army, Uganda*. Somerville, MA: Feinstein International Center.

57 Riley, Heidi. 2019. "Male Collective Identity in the People's Liberation Army of Nepal." *International Feminist Journal of Politics* 21(4): 544–565.

58 Dietrich Ortega, Luisa Maria. 2012. "Looking Beyond Violent Militarized Masculinities." *International Feminist Journal of Politics* 14(4): 489–507, 493–497; Wood, "Armed Groups and Sexual Violence," 135–136.

interpret these disparate findings? First, wartime rape by state troops with women combatants frequently receives extensive coverage from media outlets and human rights advocacy groups, considering how women affiliated with state militaries may become passive bystanders or active perpetrators of such sexual predation. Female perpetration of rape and sexual torture during conflict "often becomes the central focus of attention when reporting these atrocities" due to the "shock factor associated with [such] a relatively rare phenomenon,"[59] and while other women personnel are portrayed as war heroines, such "conflicting images of women populate news headlines and political discourses."[60] Second, rebel groups that incorporate both women combatants and leaders, despite certain glass ceilings, tend to capitalize on gender framing to propagate their public commitment and legitimacy and hence can suppress claims of rape.[61] In addition, Western reports on female insurgents and commanders sometimes expunge "aspects that might come across as controversial" to glorify their struggle.[62]

The findings from my analysis have important implications for the WPS agenda and broadly conceived military operations as they relate to combating the recalcitrant problem of wartime rape. Despite some important advancements made through the WPS agenda in developing state-level policies that prevent and prosecute sexual crimes, liberal feminism has been criticized for failing to impose substantive challenges to the gendered power relations in the broader military communities.[63] In most countries, women's military roles in public armies remain distant from the frontlines, and their proportional representation is low in absolute terms. Even in the public militaries of Israel, the United States, and the United Kingdom, where women are allowed to serve in combat units,[64] "institutional and operational requirements" often force a tokenistic inclusion of women regardless of men's gendered interests, thereby re-inscribing traditional

59 Cohen, Hoover Green, and Wood, *Wartime Sexual Violence*; Smeulers, Alette. 2015. "Female Perpetrators: Ordinary or Extra-Ordinary Women?" *International Criminal Law Review* 15(2): 207–253, 230.

60 Sjoberg, "Agency, Militarized Femininity and Enemy Others," 82.

61 Alison, Miranda H. 2009. *Women and Political Violence: Female Combatants in Ethno-National Conflict*. London, UK: Routledge; Henshaw *et al.*, "Understanding Women at War."

62 Toivanen, Mari, and Bahar Baser. 2016. "Gender in the Representations of an Armed Conflict: Female Kurdish Combatants in French and British Media." *Middle East Journal of Culture and Communication* 9(3): 294–314.

63 Heathcote, "Security Council Resolution 2242 on Women, Peace, and Security."

64 The data used in this chapter does not reflect the changes to women's engagement in combat in the latter two countries over the last few years.

gender norms on women personnel.[65] Arguably, the most important factor that has prevented state militaries from becoming fully accessible to women is "traditional, conservative, and in some ways 'macho' culture."[66] The initiatives to increase the number of women in military ranks strengthen the formal equality that aligns with liberal feminist objectives. Undoubtedly, however, there is still room to broaden the feminist project to adequately consider the factors that propel women's military service toward genuine equality and socialization of both genders, reducing wartime rape as opportunistic behaviors and heterosexual bonding practices. It is critical for the WPS imperatives and state-level policies to pay more attention to opportunity structures and other gendered social and organizational factors that women face. Otherwise, liberal feminist ideals of incorporating substantial proportions of women are unlikely to be materialized in the first place, and even with tokenistic representation, it is less likely to transform "endemic military cultures and social and power relations" in state militaries and as a result, wartime rape perpetration.[67]

Rebel organizations, with more meaningful integration of women combatants, appear to achieve what liberal feminists wish for through the increased representation of women in state militaries. Women join rebel groups for various personal, strategic, and structural reasons during times of conflict,[68] and rebel organizations have instrumental reasons to incorporate substantial proportions of women in their ranks, for example, for suicide operations, inclusive recruitment pool and ideological objectives, public relations and media attention, combat morale and organizational

65 King, "The Female Combat Soldier," 129.

66 Robinson, Lori, and Michael E. O'Hanlon. 2020. "*Women Warriors: The Ongoing Story of Integrating and Diversifying the American Armed Forces*." Brookings.

67 Ju, Changwook. 2022. "Why Do Military Officers Condone Sexual Violence? A General Theory of Commander Tolerance." *International Studies Review* 24(3): 1-31, 21.

68 Bloom, Mia. 2011. *Bombshell: The Many Faces of Women Terrorists*. Toronto, Canada: Viking Canada; Lobao, Linda M. 1990. "Women in Revolutionary Movements: Changing Patterns of Latin American Guerrilla Struggle." *Dialectical Anthropology* 15(2/3): 211–232; Lyons, Tanya. 2004. *Guns and Guerilla Girls: Women in the Zimbabwean National Liberation Struggle*. Trenton, NJ: Africa World Press; Molyneux, Maxine. 1985. "Mobilization without Emancipation? Women's Interests, the State, and Revolution in Nicaragua." *Feminist Studies* 11(2): 227–254; Oriola, Temitope. 2016. "'I Acted Like a Man': Exploring Female Ex-Insurgents' Narratives on Nigeria's Oil Insurgency." *Review of African Political Economy* 43(149): 451–469; Parkinson, Sarah Elizabeth. 2013. "Organizing Rebellion: Rethinking High-Risk Mobilization and Social Networks in War." *American Political Science Review* 107(3): 418–432; Viterna, Jocelyn. 2013. *Women in War: The Micro-Processes of Mobilization in El Salvador*. New York, NY: Oxford University Press.

stability, and demoralization of the enemy.[69] Notably, the motivations of women fighters and rebel organizations frequently coincide with rising awareness of feminism in opposition to patriarchalism and structural gender inequality.[70] Women's substantial representation in the ranks of the Liberation Tigers of Tamil Eelam in Sri Lanka and the Revolutionary Armed Forces of Colombia, where female combatants hold comparable responsibilities to male comrades-in-arms and view combat as a means of earning respect regardless of gender, provides evidence that women often join rebel groups for gender equality and emancipation, attributes that the groups are willing to encompass strategically.[71] When women's substantial inclusion is influenced by these congruent decisions made by women and rebel organizations alike,[72] collective rape as part of opportunistic violence and as a means of communicating brutal, heterosexual masculine norms is less likely to be tolerated and practiced.

Conclusion

Debates persist as to whether and how the presence of women combatants influences an armed group's proclivity for rape. Liberal feminists support the inclusion of women personnel in militaries to combat gendered norms and violent practices and hence rape during wartime. On the contrary, anti-militarist feminists paint a sobering picture, arguing that women's incorporation into militaries would reinforce heterosexual, militarized masculinity and persist belligerents' wartime rape tendencies. Non-feminist academics have also documented settings in which women combatants do

---

69 Alison, *Women and Political Violence;* Herrera, Natalia, and Douglas Porch. 2008. "'Like Going to a Fiesta' – the Role of Female Fighters in Colombia's FARC-EP." *Small Wars & Insurgencies* 19(4): 609–634; Lobao, "Women and Revolutionary Movements"; Stone, Jennie, and Katherine Pattillo. 2011. "Al-Qaeda's Use of Female Suicide Bombers in Iraq: A Case Study." In *Women, Gender, and Terrorism,* eds. Laura Sjoberg and Caron E. Gentry. Athens, OH: University of Georgia Press, 83–95.

70 International Labour Office. 2006. *Red Shoes: Experiences of Girl-Combatants in Liberia.* Geneva, Switzerland: International Labour Office; Kampwirth, Karen. 2004. *Feminism and the Legacy of Revolution: Nicaragua, El Salvador, Chiapas.* 1st edition. Athens, OH: Ohio University Press; Lobao, "Women in Revolutionary Movements"; Reif, Linda L. 1986. "Women in Latin American Guerrilla Movements: A Comparative Perspective." *Comparative Politics* 18(2): 147–169; Wilson, Amrit. 1991. *The Challenge Road: Women in the Eritrean Revolution.* London, UK: Routledge.

71 Alison, *Women in Political Violence;* Hererra and Porch, "'Like Going to A Fiesta.'"

72 Donnelly, Phoebe. 2018. "The Interactive Relationship between Gender and Strategy." *Global Society* 32(4): 457–476.

or do not moderate rape during conflict. Existing empirical analyses have come to emulate these inconclusive theoretical predictions, generating mixed and unforeseen results regarding women's combat involvement and wartime rape by belligerents.

This chapter resolves these inconclusive results and offers differential implications of female combatancy for wartime rape by public and rebel armies. That is, the presence of women combatants exacerbates wartime rape by state forces but dampens that by rebel groups. These findings intimate systemic, long-standing gendered hierarchies in state militaries, while also highlighting the possibility for women to pacify masculine bonding and misogynistic practices in rebel organizations. In short, the findings on state armies support the anti-militarist feminist view on women's presence and state-directed wartime rape,[73] while those on rebel organizations support the intuitions advocated by prominent theorists regarding women's presence and rebel-perpetrated rape.

The findings are grounded in existing data on women's representation in militaries and wartime rape that runs through 2009. Therefore, it is important to recognize that different patterns may emerge with new data collection. Nonetheless, this chapter offers some vital lessons. The way the WPS agenda and state military policies approach women's participation in militaries and wartime rape must be bolstered to better reflect the factors that perpetuate gendered dynamics and deter women's equal and meaningful engagement in military operations. Otherwise, simply increasing gender representation would face endemic resistance and be ineffective or counterproductive in combatting/not combatting wartime rape.

It is thus noteworthy that the interplay between women's motivations to fight with rebel groups and rebel leadership's governance strategy to boost female representation transpire in tandem with the clear restraint of rebel-perpetrated rape. Women in female-intensive insurgencies are treated equally while earning considerable respect and appreciation from their comrades, irrespective of gender. The women simultaneously provide strategic and tactical advantages to the rebel groups, making rape unlikely to occur as a result of opportunism or violent social dynamics. Future WPS

73 This does not mean that the liberal feminist stance on women's combat participation and wartime rape is misguided. The principal condition under which liberal feminist ideas are materialized is a sufficiently strong representation of women in the diverse ranks of militaries. See, for example, Card's "Rape as a Weapon of War," 12. The findings in this chapter therefore indicate that such strong representation is exceedingly difficult to meet in state armies.

policy initiatives should thus take cues from insurgent women,[74] focusing on such conditions under which women's presence transforms gendered norms and violent practices in military operations.

References

Alison, Miranda H. 2009. *Women and Political Violence: Female Combatants in Ethno-National Conflict*. London, UK: Routledge.

Bagozzi, Benjamin E. 2015. "Forecasting Civil Conflict with Zero-Inflated Count Models." *Civil Wars* 17(1): 1–24.

Bagozzi, Benjamin E., Daniel W. Hill, Will H. Moore, and Bumba Mukherjee. 2015. "Modeling Two Types of Peace: The Zero-Inflated Ordered Probit (ZiOP) Model in Conflict Research." *Journal of Conflict Resolution* 59(4): 728–752.

Beckwith, Karen. 2007. "Numbers and Newness: The Descriptive and Substantive Representation of Women." *Canadian Journal of Political Science* 40(1): 27–49.

Bloom, Mia. 2011. *Bombshell: The Many Faces of Women Terrorists*. Toronto, Canada: Viking Canada.

Bridges, Donna, and Debbie Horsfall. 2009. "Increasing Operational Effectiveness in UN Peacekeeping: Toward a Gender-Balanced Force." *Armed Forces & Society* 36(1): 120–130.

Bystydzienski, Jill M. 1993. "Women in Groups and Organizations: Implications for the Use of Force." In *Women and the Use of Military Force*, eds. Ruth H. Howes and Michael R. Stevenson. Boulder, CO: Lynne Rienner Publishers, 39–52.

Card, Claudia. 1996. "Rape as a Weapon of War." *Hypatia* 11(4): 5–18.

Carlson, Kristopher, and Dyan Mazurana. 2008. *Forced Marriage within the Lord's Resistance Army, Uganda*. Somerville, MA: Feinstein International Center.

Cohen, Dara Kay, Amelia Hoover Green, and Elisabeth Jean Wood. 2013. *Wartime Sexual Violence: Misconceptions, Implications, and Ways Forward*. Washington, DC: United States Institute of Peace.

Cohen, Dara Kay, and Amelia Hoover Green. 2012. "Dueling Incentives: Sexual Violence in Liberia and the Politics of Human Rights Advocacy." *Journal of Peace Research* 49(3): 445–458.

Cohen, Dara Kay, and Ragnhild Nordås. 2014. "Sexual Violence in Armed Conflict: Introducing the SVAC Dataset, 1989–2009." *Journal of Peace Research* 51(3): 418–428.

Cohen, Dara Kay. 2013. "Female Combatants and the Perpetration of Violence: Wartime Rape in the Sierra Leone Civil War." *World Politics* 65(3): 383–415.

Cohen, Dara Kay. 2016. *Rape during Civil War*. Ithaca, NY: Cornell University Press.

74 For a helpful discussion on why women in rebel organizations may receive little attention in the WPS agenda, see Henshaw, Alexis Leanna. 2017. *Making Violent Women Visible in the WPS Agenda*. London, UK: LSE Centre for Women, Peace and Security.

Defeis, Elizabeth F. 2008. "U.N. Peacekeepers and Sexual Abuse and Exploitation: An End to Impunity." *Washington University Global Studies Law Review* 7(2): 185–214.

DeGroot, Gerard J. 2001. "A Few Good Women: Gender Stereotypes, the Military and Peacekeeping." *International Peacekeeping* 8(2): 23–38.

Desmarais, Bruce A., and Jeffrey J. Harden. 2013. "Testing for Zero Inflation in Count Models: Bias Correction for the Vuong Test." *Stata Journal* 13(4): 810–835.

Dietrich Ortega, Luisa Maria. 2012. "Looking Beyond Violent Militarized Masculinities." *International Feminist Journal of Politics* 14(4): 489–507.

Donnelly, Phoebe. 2018. "The Interactive Relationship between Gender and Strategy." *Global Society* 32(4): 457–476.

Duncanson, Claire, and Rachel Woodward. 2016. "Regendering the Military: Theorizing Women's Military Participation." *Security Dialogue* 47(1): 3–21.

Enloe, Cynthia H. 2000. *Maneuvers: The International Politics of Militarizing Women's Lives*. Berkeley, CA: University of California Press.

Goldstein, Joshua S. 2001. *War and Gender: How Gender Shapes the War System and Vice Versa*. Cambridge: Cambridge University Press.

Haer, Roos, and Tobias Böhmelt. 2018. "Girl Soldiering in Rebel Groups, 1989–2013: Introducing a New Dataset." *Journal of Peace Research* 55(3): 395–403.

Harris, Mark N., and Xueyan Zhao. 2007. "A Zero-Inflated Ordered Probit Model, with an Application to Modelling Tobacco Consumption." *Journal of Econometrics* 141(2): 1073–1099.

Heathcote, Gina. 2018. "Security Council Resolution 2242 on Women, Peace and Security: Progressive Gains or Dangerous Development?" *Global Society* 32(4): 374–394.

Henshaw, Alexis *et al.* 2019. "Understanding Women at War: A Mixed-Methods Exploration of Leadership in Non-State Armed Groups." *Small Wars & Insurgencies* 30(6–7): 1089–1116.

Henshaw, Alexis Leanna. 2016. "Where Women Rebel: Patterns of Women's Participation in Armed Rebel Groups 1990–2008." *International Feminist Journal of Politics* 18(1): 39–60.

Henshaw, Alexis Leanna. 2017. *Making Violent Women Visible in the WPS Agenda*. London, UK: LSE Centre for Women, Peace and Security.

Herrera, Natalia, and Douglas Porch. 2008. "'Like Going to a Fiesta' – the Role of Female Fighters in Colombia's FARC-EP." *Small Wars & Insurgencies* 19(4): 609–634.

Hoover Green, Amelia. 2017. "Armed Group Institutions and Combatant Socialization: Evidence from El Salvador." *Journal of Peace Research* 54(5): 687–700.

Hoover Green, Amelia. 2018. *The Commander's Dilemma: Violence and Restraint in Wartime*. Ithaca, NY: Cornell University Press.

International Labour Office. 2006. *Red Shoes: Experiences of Girl-Combatants in Liberia*. Geneva, Switzerland: International Labour Office.

Ju, Changwook. 2022. "Why Do Military Officers Condone Sexual Violence? A General Theory of Commander Tolerance." *International Studies Review* 24(3): 1–31.

Kampwirth, Karen. 2004. *Feminism and the Legacy of Revolution: Nicaragua, El Salvador, Chiapas*. 1st edition. Athens, OH: Ohio University Press.

Karim, Sabrina, and Kyle Beardsley. 2016. "Explaining Sexual Exploitation and Abuse in Peacekeeping Missions: The Role of Female Peacekeepers and Gender Equality in Contributing Countries." *Journal of Peace Research* 53(1): 100–115.

King, Anthony. 2016. "The Female Combat Soldier." *European Journal of International Relations* 22(1): 122–143.

Kirby, Paul, and Laura J. Shepherd. 2016. "The Futures Past of the Women, Peace and Security Agenda." *International Affairs* 92(2): 373–392.

Kramer, Sophie. 2012. "Forced Marriage and the Absence of Gang Rape: Explaining Sexual Violence by the Lord's Resistance Army in Northern Uganda." *Journal of Politics and Society* 23(1): 11–49.

Kronsell, Annica. 2005. "Gendered Practices in Institutions of Hegemonic Masculinity." *International Feminist Journal of Politics* 7(2): 280–298.

Krüger, Jule, and Ragnhild Nordås. 2020. "A Latent Variable Approach to Measuring Wartime Sexual Violence." *Journal of Peace Research* 57(6): 728–739.

Leiby, Michele. 2009. "Digging in the Archives: The Promise and Perils of Primary Documents." *Politics & Society* 37(1): 75–99.

Leiby, Michele L. 2012. "The Promise and Peril of Primary Documents: Documenting Wartime Sexual Violence in El Salvador and Peru." In *Understanding and Proving International Sex Crimes*, eds. Morten Bergsmo, Alf B. Skre, and Elisabeth Jean Wood. Beijing: Torkel Opsahl Academic EPublisher, 315–366.

Lobao, Linda M. 1990. "Women in Revolutionary Movements: Changing Patterns of Latin American Guerrilla Struggle." *Dialectical Anthropology* 15(2/3): 211–232.

Loken, Meredith. 2017. "Rethinking Rape: The Role of Women in Wartime Violence." *Security Studies* 26(1): 60–92.

Lyons, Tanya. 2004. *Guns and Guerilla Girls: Women in the Zimbabwean National Liberation Struggle*. Trenton, NJ: Africa World Press.

MacKinnon, Catharine A. 2006. *Are Women Human? And Other International Dialogues*. Cambridge, MA: Belknap Press of Harvard University Press.

Mehrl, Marius. 2022. "Female Combatants and Wartime Rape: Reconsidering the Role of Women in Armed Conflict." *Armed Forces & Society* 48(2): 464–479.

Mobekk, Eirin. 2010. "Gender, Women and Security Sector Reform." *International Peacekeeping* 17(2): 278–291.

Molyneux, Maxine. 1985. "Mobilization without Emancipation? Women's Interests, the State, and Revolution in Nicaragua." *Feminist Studies* 11(2): 227–254.

Oriola, Temitope. 2016. "'I Acted Like a Man': Exploring Female Ex-Insurgents' Narratives on Nigeria's Oil Insurgency." *Review of African Political Economy* 43(149): 451–469.

Parkinson, Sarah Elizabeth. 2013. "Organizing Rebellion: Rethinking High-Risk Mobilization and Social Networks in War." *American Political Science Review* 107(3): 418–432.

Reardon, Betty. 1985. *Sexism and the War System*. New York: Teachers College Press.

Reif, Linda L. 1986. "Women in Latin American Guerrilla Movements: A Comparative Perspective." *Comparative Politics* 18(2): 147–169.

Riley, Heidi. 2019. "Male Collective Identity in the People's Liberation Army of Nepal." *International Feminist Journal of Politics* 21(4): 544–565.

Robinson, Lori, and Michael E. O'Hanlon. 2020. "Women Warriors: The Ongoing Story of Integrating and Diversifying the American Armed Forces." *Brookings.*

Roth, Françoise, Tamy Guberek, and Amelia Hoover Green. 2011. *Using Quantitative Data to Assess Conflict-Related Sexual Violence in Colombia: Challenges and Opportunities.* Benetech Human Rights Program and Corporación Punto de Vista.

Shepherd, Laura J. 2007. "'Victims, Perpetrators and Actors' Revisited: Exploring the Potential for a Feminist Reconceptualisation of (International) Security and (Gender) Violence." *British Journal of Politics and International Relations* 9(2): 239–256.

Simić, Olivera. 2010. "Does the Presence of Women Really Matter? Towards Combating Male Sexual Violence in Peacekeeping Operations." *International Peacekeeping* 17(2): 188–199.

Sjoberg, Laura, and Sandra Via. 2010. *Gender, War, and Militarism Feminist Perspectives.* Santa Barbara, CA: Praeger.

Sjoberg, Laura. 2007. "Agency, Militarized Femininity and Enemy Others: Observations from the War in Iraq." *International Feminist Journal of Politics* 9(1): 82–101.

Sjoberg, Laura. 2016. *Women as Wartime Rapists: Beyond Sensation and Stereotyping.* New York: New York University Press.

Smeulers, Alette. 2015. "Female Perpetrators: Ordinary or Extra-Ordinary Women?" International Criminal Law Review 15(2): 207–253.

Stiehm, Judith Hicks, ed. 1996. *It's Our Military, Too! Women and the U.S. Military.* Philadelphia, PA: Temple University Press.

Stone, Jennie, and Katherine Pattillo. 2011. "Al-Qaeda's Use of Female Suicide Bombers in Iraq: A Case Study." In *Women, Gender, and Terrorism,* eds. Laura Sjoberg and Caron E. Gentry. Athens, OH: University of Georgia Press, 83–95.

Thomas, Jakana L., and Kanisha D. Bond. 2015. "Women's Participation in Violent Political Organizations." *American Political Science Review* 109(3): 488–506.

Thomas, Jakana L, and Reed M Wood. 2018. "The Social Origins of Female Combatants." *Conflict Management and Peace Science* 35(3): 215–232.

Toivanen, Mari, and Bahar Baser. 2016. "Gender in the Representations of an Armed Conflict: Female Kurdish Combatants in French and British Media." Middle East Journal of Culture and Communication 9(3): 294–314.

Valenius, Johanna. 2007. "A Few Kind Women: Gender Essentialism and Nordic Peacekeeping Operations." *International Peacekeeping* 14(4): 510–523.

Viterna, Jocelyn. 2013. *Women in War: The Micro-Processes of Mobilization in El Salvador.* New York, NY: Oxford University Press.

Vuong, Quang H. 1989. "Likelihood Ratio Tests for Model Selection and Non-Nested Hypotheses." *Econometrica* 57(2): 307–333.

Whitworth, Sandra. 2004. *Men, Militarism, and UN Peacekeeping: A Gendered Analysis.* Boulder, CO: Lynne Rienner.

Wilson, Amrit. 1991. *The Challenge Road: Women in the Eritrean Revolution.* London, UK: Routledge.

Wilson, Paul. 2015. "The Misuse of the Vuong Test for Non-Nested Models to Test for Zero-Inflation." *Economics Letters* 127(C): 51–53.

Wood, Elisabeth Jean. 2006. "Variation in Sexual Violence During War." *Politics & Society* 34(3): 307–342.

Wood, Elisabeth Jean. 2009. "Armed Groups and Sexual Violence: When Is Wartime Rape Rare?" *Politics & Society* 37(1): 131–161.

Wood, Reed M, and Jakana L Thomas. 2017. "Women on the Frontline: Rebel Group Ideology and Women's Participation in Violent Rebellion." *Journal of Peace Research* 54(1): 31–46.

Wooldridge, Jeffrey M. 2013. *Introductory Econometrics: A Modern Approach*. 5th edition. Mason, OH: Cengage Learning.

6

ENHANCING OPERATIONAL EFFECTIVENESS

Through an Analysis of the Women, Peace and Security Environment in Djibouti

Whitney Grespin

Djibouti is a focal point of geopolitical competition in the Horn of Africa. Its strategic location, abutting one of the world's busiest shipping lanes and acting as the terminus of inland rail traffic from Ethiopia, give it significant global importance in this era of increasing great power competition. That it is also host to the United States' only permanent military base in Africa is foundational to interest of the U.S. Government.[1]

In 2000, the United Nations Security Council adopted Resolution 1325 (UNSCR 1325) on Women, Peace and Security. UNSCR 1325 asserts that peace and security efforts are more durable when women are treated as equal partners in interventions to prevent violent conflict, deliver relief and recovery, and promote lasting peace. The development of government-led National Action Plans (NAPs) and other national level strategies has been undertaken since 2005. NAPs are foundational to the operationalization of the aims of the Women, Peace and Security (WPS) agenda and allow states to design their own context-specific execution of measures to mitigate the marginalization of vulnerable populations. However, a paucity of timely information gathering and uniform reporting on NAP outcomes has left significant gaps in the body of knowledge about progress towards these goals.

1 Central Intelligence Agency, "CIA World Factbook: Djibouti."

Djibouti faces complex challenges when it comes to realizing the WPS agenda. Education and employment are among the most prominent. A reported 60.5 per cent of Djiboutian women are illiterate, with only 19 per cent of women employed (in comparison to 81 per cent of men), thus stunting the potential of Djiboutian women who suffer from the related effects of poverty.[2] Djiboutian youth unemployment is among the highest in the world at 70 per cent and this also disproportionately affects women and girls. The widespread practice of female genital mutilation (FGM) reflects Djibouti's lack of gender equality. A 1995 law prohibiting the practice has never been enforced. FGM is also a major contributor to high rates of maternal and infant mortality and other obstetrical complications.[3] While Djiboutian narratives maintain that citizens are equal before the law, it is evident that gender equality, in terms of both opportunities and outcomes, is far from realized.[4]

Djibouti's NAP originated in November 2017. It was scoped through 2020 and is currently under revision. It serves as a binding national document for the integration of WPS principles, including how they will be governed and prioritized within Djiboutian institutions. The NAP is intended to assist in identifying and resourcing priorities, determining responsibilities, and committing to action.[5] The 2017 NAP sought to address four main issues: "the situation of women and sexual violence at the national level, refugee women and sexual violence, women in pastoral conflicts, and the issues of sexual violence regarding security and defense forces."[6] It addresses these through engagement with the three Ps of the WPS agenda:[7]

- **Participation** of women in all levels of decision-making for the prevention, management, and regulation of conflicts,
- **Prevention** of gender-based violence, sexual abuse, and the exploitation of women and girls, and
- **Protection** and security of sexual violence victims during the conflict.

---

2 Ibid

3 Ibid.

4 Bertelsmann Stiftung, "2022 Djibouti Country Report," 26.

5 The full implementation of the NAP was estimated to require $913,000, "including expenses for communication, coordination, monitoring and evaluation, and capacity building" for the Ministry of Women and Family, which was designated to have primacy over the NAP. Ministère de la Femme et de la Famille, *Plan d'Action National pour la Mise en œuvre de la Résolution 1325 (2000) et des Resolutions Connexes du Conseil de Sécurité des Nations Unies sur la Femme et la Paix.*

6 Ibid

7 Ibid.

Thirteen additional objectives were also established to improve the protection and promotion of women's rights during conflicts and/or post-conflict.[8] Although it was reported that 42 indicators were proposed to review the implementation of the NAP (each of which referred to a proposed action), information regarding the execution of proposed actions is not readily available in the public domain. While this does not necessarily mean that actions relevant to the 42 indicators have not been undertaken in internal Djiboutian government activity that has not been made publicly available, the paucity of available data in the context of the Djiboutian NAP raises the prospect of investigating the utility of alternative tools. The core of this chapter demonstrates the application of a U.S. Department of Defense analytical tool, PMESII-PT, augmented by an overlay of gender considerations can provide an expanded understanding of the operational environment.

This chapter will illustrate the value of broadening the aperture of the existing PMESII-PT tool (described below) to improve partner nations' understanding of the operational environment. It increases situational awareness and capacity to assist meaningfully and appropriately in realizing WPS objectives. The PMESII-PT framework will be applied to the Djiboutian context to provide an impression of the impacts of efforts to operationalize the WPS agenda and the outcomes of the Djiboutian NAP to date. It will then capture actionable recommendations distilled from a contemporary case study to identify best practices in complex environments and to support the pursuit of NAP priorities in both Djiboutian and broader contexts. As these challenges are not necessarily unique to the Djiboutian context, an evaluation of WPS-related efforts and best practices guided by Djibouti's NAP might prove informative to policymakers and practitioners seeking similar aims elsewhere.

Using PMESII-PT in Gender Analysis

Scholarship on WPS is increasingly focused on NAP efficacy[9] as being foundational to the broader WPS agenda,[10] but highlights the paucity of data

---

8 Ibid.

9 Hamilton, Naam, and Shepherd, *Twenty Years of Women, Peace and Security National Action Plans: Analysis and Lessons Learned*.

10 Jacevic, "WPS, States, and the National Action Plans," in *The Oxford Handbook of Women, Peace, and Security*, edited by Sara E Davies and Jacqui True, 273-90. Oxford: Oxford University Press, 2019.

available to accurately evaluate progress.[11] In lieu of referencing consistent, reliable, and systematically collected data, an environmental assessment of the operating environment can be conducted using the analytical tool PMESII-PT. PMESII-PT is an acronym developed by the U.S. Army which stands for Political, Military, Economic, Social, Information, Infrastructure, Physical Environment, and Time. The tool is used to organize information to inform strategy development while preparing for and executing operations in foreign countries, particularly in regard to mission analysis and course of action (CoA) development. It can also be useful in gauging the overall progress towards priorities articulated in NAPs and understanding the windows of opportunity for positive engagement to further the four pillars of the WPS: participation, prevention, protection, and relief and recovery. Building upon the general situational awareness provided using a PMESII-PT framework is the explicit examination of gender dynamics through a "gender analysis."

The Canadian government defines gender analysis as a tool that is:

> Used to understand the relationships between men and women, their access to resources, their activities, and the constraints they face relative to each other. Gender analysis provides information that recognizes that gender, and its relationship with race, ethnicity, culture, class, age, disability, and/or other status, is important in understanding the different patterns of involvement, behaviour and activities that women and men have in economic, social and legal structures.[12]

Gender analysis highlights the varied roles that men, women, boys, and girls play in family, community, economic, legal, and political structures in different contexts (regions and/or countries). It can provide insights for the different circumstances that women, men, girls, and boys face, and the disparate effects that policies and programs might have on them. Examining this information can ensure that military operations do not exacerbate conditions on the ground. Gender awareness or employing a "gender perspective" highlights the reasons for the current division of responsibilities and benefits, and their effect(s) on incentives and the

11 While beyond the remit of this chapter, such scholarship is important to understanding how differing interpretations of the pillars' intents might affect holistic monitoring and evaluation of progress towards the WPS agenda.

12 Government of Canada. "Gender Analysis." April 23, 2019. https://www.international.gc.ca/world-monde/funding-financement/gender_analysis-analyse_comparative.aspx?lang=eng.

distribution of rewards. Gender analyses are important as they inform and improve policies and programs designed to ensure that their needs are considered. They are illustrative of the gender mainstreaming that UNSCR 1325 advocates in member state programming.

In August 2022, a practitioner and Atlantic Council Nonresident Senior Fellow Cori Fleser, published a gender analysis of Ukraine.[13] Her work highlights how gender analysis can provide militaries with a more comprehensive understanding and appreciation of a range of information about the differing ways in which war impacts civilians.[14] This might be based on social norms and behaviors associated with their identity as male, female, or non-binary alongside other identity characteristics such as disability status, ethnicity, or religion. Fleser is clear that the piece "does not argue that gender analysis can or should be used to wage war more effectively from a purely military operational perspective. It takes the position that to be considered militarily effective, military action must meet its desired objectives while mitigating cost and negative impact on civilians and humanitarian actors operating in the same space."[15] She points out that although it may seem contradictory to discuss the risks faced by women and girls in this conflict and then highlight their agency, it is not, in fact, dichotomous. The complexity that gender analysis illuminates, and the nuance it demands of either humanitarian or military interventions in complex environments requires the elimination of wholesale assumptions about women's victimization, and instead calls for an examination of motivations as a critical component of understanding gender dynamics in conflict.

Fleser's gender analysis for Ukrainian security assistance lays out several core recommendations to guide interventions to accommodate gendered human security challenges, and notes that "they are not all encompassing and should be reevaluated as the [context] evolves and new information becomes available."[16] They are divided into tactical, operational, and strategic recommendations.

---

13 Fleser, Cori. "Beyond Munitions: A Gender Analysis for Ukrainian Security Assistance."
14 Fleser's work builds on the UN Women and CARE International's Rapid Gender Analysis of Ukraine. See https://www.unwomen.org/en/digital-library/publications/2022/05/rapid-gender-analysis-of-ukraine
15 Fleser, Cori. "Beyond Munitions: A Gender Analysis for Ukrainian Security Assistance."
16 Ibid.

Tactical:

1. Incorporate human security guidance into commander's intent.
2. Develop and provide specialized training for deployed forces.
3. Balance threat and civilian intelligence assessments.

Operational:

1. Provide gender-specific healthcare and equipment.
2. Counter gender disinformation.
3. Deploy civil-military cooperation capabilities.
4. Diversify force composition for deployed CIMIC teams.

Strategic:

1. Bolster [intervener's] internal gender and human security capacity.
2. Advocate for women's participation in peace negotiations.
3. Elevate the protection of civilians [in discourse and negotiations].

Although the recommendations are geared towards a military audience, they can be easily adapted for a variety of actors and audiences to promote gender mainstreamed programming and operational effectiveness. They provide concrete guidance on how gaps identified in the PMESII-PT analysis can be proactively addressed when designing interventions or planning assistance and feed into the **prevention** pillar of WPS.

Below is a sample analysis of the operational environment in Djibouti using the PMESII-PT model, to include a dynamic that, while brief, is representative of the type of consideration of gender dimensions that would allow a more holistic perspective of how the environment might be more fully understood. The use of a gendered lens highlights broader considerations that otherwise might not necessarily be identified as threats or opportunities for increased engagement or stabilization.

Political

In Djibouti both parliamentary and presidential elections are held regularly, "but usually hold little surprise."[17] As Freedom House's landing page for Djibouti summarizes,

17 Bertelsmann Stiftung, "2022 Djibouti Country Report," 10.

> Djibouti is a republic ruled by a powerful president, Ismail Omar Guelleh, who has been in office since 1999 and is not subject to term limits. While Djibouti technically has a multiparty political system, the ruling Union for a Presidential Majority (UMP) uses authoritarian means to maintain its dominant position. The opposition's ability to operate is severely constrained, and journalists and activists who air criticism of Guelleh or the UMP are regularly harassed or arrested.[18]

The Fragile States Index highlights significant human rights issues including:[19]

- Unlawful or arbitrary killings including extrajudicial killings;
- Cases of cruel, inhuman, or degrading treatment or punishment by the government;
- Harsh and life-threatening prison conditions;
- Arbitrary arrest or detention;
- Arbitrary or unlawful interference with privacy;
- Serious restrictions on free expression, the press, and the internet, including violence, threats of violence, or unjustified arrests or prosecutions against journalists, censorship, site blocking, and the existence of criminal libel laws;
- Substantial interference with the freedom of peaceful assembly and freedom of association; and
- The worst forms of child labor committed throughout the country.

Djibouti has seen negative trends in the Transparency International's Corruption Perception Index[20] over the last decade.[21] Corruption and the misappropriation of funds is a major problem, neither the state budget nor Inspector General reports are publicly available.

A 2022 country report on Djibouti emphasizes this corruption as present at all levels of public administration,[22] and a barrier to both political and economic participation and, thus, representation. It further highlights how the government's domination of public affairs ensures that segments of society remain overall underrepresented and generally structurally

18 Freedom House, "Freedom in the World: Djibouti Country Report."
19 Fund for Peace, "Fragile States Index: Country Dashboard Djibouti."
20 The CPI ranks 180 countries and territories around the world by their perceived levels of public sector corruption. The results are given on a scale of 0 (highly corrupt) to 100 (very clean). Djibouti currently has a score of 30/100. Transparency International, "Corruption Perceptions Index."
21 Transparency International, "Corruption Perceptions Index."
22 Bertelsmann Stiftung, "2022 Djibouti Country Report," 14.

weak, noting that, "while organizations focusing on issues related to socioeconomic development can operate relatively freely, groups advocating for politically sensitive topics (for example, human rights) usually face difficulties in registering and are reportedly subject to unexpected audits by the government."[23] While this does not explicitly flag gender inequality as politically sensitive, there is implicit evidence across the remainder of this PMESII-PT analysis that there is insufficient consideration of the disproportionate impact of hardships on specific population segments.

Although women have had the vote in Djibouti since 1946, it was forty years later that female candidates were permitted to run for office. The first female member of parliament was elected in the 2003 elections. A legal quota now ensures that women hold at least a quarter of the seats in the National Assembly, however, the current number is only slightly above that level. As summarized by Freedom House, "In practice, the authoritarian political system restricts the ability of women and members of ethnic and religious minorities to organize independently and advance their respective interests."[24] Although consistent data is unavailable to surveille the full context, it has been reported that "nearly half of all judges in the country are female and a few ministries are headed by women,"[25] but also that "women remain under-represented in senior government positions and the economy is largely dominated by men."[26]

Military

As of 2021, China,[27] France, Italy, Japan, and the US maintain bases[28] in Djibouti for regional military missions such as crisis response, counter-piracy, counterterrorism, and security assistance. As the former colonial power, France has multiple bases and hosts troop contingents from Germany and Spain, while the EU and NATO have also maintained a presence to support maritime training efforts and multinational naval

23 Ibid 18.
24 Freedom House, "Freedom in the World: Djibouti Country Report."
25 Ibid 19.
26 Ibid 26.
27 "Security and economic ties with Beijing are particularly strong… Recently, China has become a major partner of the government in infrastructure development and financing." Bertelsmann Stiftung, "2022 Djibouti Country Report," 33.
28 The CIA World Factbook publication reports that in 2017 Djibouti and Saudi Arabia announced plans for the Saudis to build a military base there, although no start date was announced.

counter-piracy operations.[29] As the regional think tank Sahan reported in November of 2021,

> The salient feature of today's scramble for the Horn is over-militarization. Tiny Djibouti now encapsulates the risks of over-militarization.[30] There are now over 12 powers with military installations in the strategic Red Sea state... Never before in history have we seen such numbers of strategic foes, bristling with sophisticated weapons, congregated in one tiny space.[31]

Sahan provided further commentary on how, in addition to publicizing status as a consequential power, the presence of these nations served the utility of building expeditionary military capabilities and "keeping tabs on what your enemy is up to."[32]

As of 2020, the Djiboutian Armed Forces (Forces Armées Djiboutiennes [FAD]) themselves were constituted of approximately 10,500 active troops, with a breakdown of approximately 8,000 Army, 250 Naval forces, 250 Air personnel, and 2,000 Gendarmerie.[33] Djibouti is important not only to the Global North, but also within the region in terms of FAD force generation because of their role as being a core Troop Contributing Country to the AMISOM (now ATMIS) mission in Somalia. The author has encountered female members of the FAD in serendipitous interactions but was unable to access information regarding the prevalence of female employment across the Djiboutian security forces. A lack of visibility on the part of gender diversity within the armed forces is problematic.

Economic

Despite COVID-related decreases in economic activity, Djibouti had achieved relatively steady economic growth in recent years. Djibouti has pursued a model of economic development based on infrastructure through the expansion of ports and regional transportation links, capitalizing on the country's strategic location alongside the Bab-el-Mandeb Strait, the central

29 Central Intelligence Agency, "CIA World Factbook: Djibouti."
30 "Negotiations are underway for Saudi Arabia to build a naval facility. Taiwan is rapidly building its influence in the Horn and in the last year has opened a permanent mission in Somaliland. China has its first and largest overseas naval base in Djibouti, opened in 2017. Japan, an adversary of China, now has a facility in Djibouti as well." Sahan, *The Somali Wire*.
31 Sahan, *The Somali Wire*.
32 Ibid.
33 Central Intelligence Agency, "CIA World Factbook: Djibouti."

maritime corridor controlling access to the Red Sea.[34] The World Bank has assessed that Djibouti's growth prospects depend heavily on Ethiopia's political and economic conditions, as well as the progress of continued reforms to maintain the country's competitive edge in transportation.[35] The Ethiopia-Djibouti railway, which connects Ethiopia to Djibouti's ports, is particularly important to the health of the Djiboutian economy along with related foreign assistance to finance infrastructure and development projects. It has been assessed that Djibouti's growth is attributable to debt-financed investments into largescale infrastructure, posing a "considerable risk" for future growth trajectories.[36]

Djibouti's strategic trading location is home to an active trader and broker network, but as the country seeks to increase investment and promote itself as a regional financial hub it is increasingly susceptible to money laundering and other illicit financial activities. As the U.S. Department of State has noted, "The country's proximity to Somalia and Yemen remains a risk factor, as many Djibouti-based financial institutions continue to operate in neighboring countries that have weak or no AML/CFT legislation or other financial controls."[37]

While Djibouti's economy is dominated by state-owned enterprises such as agro-industries, financial services, telecommunications, and transport and storage services, it is reported that many residents have difficulty finding employment in the formal sector given that the president and his network tightly control large-scale economic activity, especially around the military bases leased by foreign powers.[38] Revenues collected from the lease of military bases to foreign nations[39] are not accounted for in the official state budget,[40] with corruption seen as remaining high and a "lack of transparency in policymaking and licensing continu[ing] to deter private sector investment."[41] Resultantly, the informal sector[42] of day

34 Bertelsmann Stiftung, "2022 Djibouti Country Report," 5.
35 World Bank, "Djibouti's Economic Update – October 2021."
36 Bertelsmann Stiftung, "2022 Djibouti Country Report," 4.
37 U.S. Department of State, "Country Reports on Terrorism: Djibouti."
38 Freedom House, "Freedom in the World: Djibouti Country Report."
39 A reported 13 per cent of Djibouti's total revenue between 2012 and 2014 came from leasing its land out to foreign militaries. International Monetary Fund, *Djibouti: Selected Issues.*
40 Bertelsmann Stiftung, "2022 Djibouti Country Report," 11.
41 Ibid 4.
42 "Given the lack of manufacturing and processing industries, employment opportunities are limited, and the growing population of working age are left with few economic opportunities. Qat (or Khat) defines daily lives in most parts of the country, given the high rates of addiction to this narcotic leaf among all segments of the populations... the proportion of available funds spent on Qat is considerable." Bertelsmann Stiftung, "2022 Djibouti Country Report," 30.

laborers, small traders, and the like plays an important role in the economy and comprises a "considerable percentage of activities both in the urban and rural areas of the country."[43]

Legal safeguards against exploitative working conditions are poorly enforced, leaving migrant workers and refugees especially vulnerable to abuse.[44] The growing influx of migrants is posing a great challenge to the Djiboutian authorities, who reportedly require additional capacity to cope with the current situation.[45] There is also evidence of the 'brothels and barracks' phenomenon, including reporting about Sexual Exploitation and Abuse (SEA) in Djibouti out of American Camp Lemonnier[46] and a likelihood of its manifestation amongst the other partner military force presence in the country.[47]

Social

Clan affiliation determines economic and, to a lesser extent, political opportunities across Djibouti.[48] Citizenship is by descent only and requires the mother to be a citizen of Djibouti.[49] It is recognized that Djiboutians rely heavily on family and clan networks for support, including remittances from members based overseas.[50] Nearly all Djiboutians (94 per cent) are Sunni Muslim, the remaining 6 per cent are a combination of Shia Muslim, Christian, Hindu, Jewish, Baha'i, and atheist.[51] Freedom House reports that, "domestic violence is rarely reported and prosecuted, and spousal rape is not specifically criminalized. The Sharia-based family code requires women to obtain a guardian's consent to marry, among other discriminatory provisions surrounding marriage and divorce."[52]

Youth unemployment (ages 15-24) is high at 73 per cent.[53] The lack of viable, continuous employment is a threat to security at both the civil society and individual levels. Women's access to income-generating activities

---

43 Bertelsmann Stiftung, "2022 Djibouti Country Report," 22.

44 Freedom House, "Freedom in the World: Djibouti Country Report."

45 UN Migration, "Djibouti."

46 Reese, "Investigating Foreign Military Bases and Sexual Violence in Djibouti."

47 Ray Acheson, *Remote Warfare and Sexual Violence in Djibouti*, https://wilpf.org/wp-content/uploads/2017/09/2017_RemoteWarfareAndSexualViolenceInDjibouti.pdf.

48 Bertelsmann Stiftung, "2022 Djibouti Country Report," 7.

49 Central Intelligence Agency, "CIA World Factbook: Djibouti."

50 International Monetary Fund, *Djibouti: Selected Issues*.

51 Central Intelligence Agency, "CIA World Factbook: Djibouti."

52 Freedom House, "Freedom in the World: Djibouti Country Report."

53 Central Intelligence Agency, "CIA World Factbook: Djibouti."

and economic participation remains limited although the Government of Djibouti has made efforts to support women's economic empowerment.[54] Areas lacking methodologies for routine monitoring include aspects of gender and poverty, physical and sexual harassment, women's access to assets (including land), and gender and the environment. As the UN Entity for Gender Equality and the Empowerment of Women (UN Women) notes, closing these gender data gaps will be essential for achieving gender-related SDG commitments in Djibouti.[55]

The Djiboutian NAP includes the participation of non-governmental organizations and feminist associations as key stakeholders in the plan's development, implementation, monitoring, and evaluation processes.[56] The Ministry for the Advancement of Women, Family Welfare and Social Affairs (Ministère délégué auprès du premier ministre chargé de la promotion de la femme, du bien-être familial et des affaires sociales, or MPF) is responsible for the protection of rights of women. There are indications that property rights are well defined in that the constitution safeguards the right to private property and a 1991 law governs land ownership, building registrations, and title deeds, but "information on the quality of the land register is not available," to include indicators regarding women's autonomous ability to control or own property.[57] The National Association of Women in Djibouti (Union Nationale des Femmes Djiboutiennes) has a membership of roughly 4,000 and is headed by the first lady.[58] While the organization is not directly under government control, the figurehead is notable, and the Association functions as a bridge between donors and non-governmental organizations to support female education and health.

Information

There are no apparent private TV or radio stations in Djibouti. The state-owned Radiodiffusion-Television de Djibouti operates the sole terrestrial TV station as well as the only two domestic radio networks.[59] The CIA

---

54 U.S. Agency for International Development, "Gender Equality and Women's Empowerment."
55 UN Women, "Djibouti."
56 Ministère de la Femme et de la Famille, *Plan d'Action National pour la Mise en œuvre de la Résolution 1325 (2000) et des Resolutions Connexes du Conseil de Sécurité des Nations Unies sur la Femme et la Paix.*
57 Bertelsmann Stiftung, "2022 Djibouti Country Report," 25.
58 Bertelsmann Stiftung, "2022 Djibouti Country Report," 37.
59 Central Intelligence Agency, "CIA World Factbook: Djibouti."

World Factbook assesses that Djibouti's poor domestic infrastructure and monopolized telecom system[60] render communications availability sparse,[61] with mobile coverage limited to Djibouti Ville.[62] Furthermore, the government reportedly slows down internet connectivity, "to restrict access to social media, which are some of the few places where freedom of speech and access to information prevail."[63] No freedom of information laws exist, and budgetary decisions are not available to the public.

The detention of journalists and "limited space granted to the [political] opposition"[64] are notable in creating an environment of information scarcity in the country. It has been reported that "there are no public opinion surveys in Djibouti," and that, "trust in the system and [its] adherence to democratic norms is low," with disapproval voiced mainly online or in private.[65] Reporters Without Borders (RSF) assesses that, "The freedom to inform is non-existent in Djibouti," and ranked the country 164/180 in its 2022 index.[66] Indeed, RSF specifically flags that, "the status of women and subjects relating to sexual orientation are particularly taboo... and are subject to censorship and self-censorship."[67] The reluctance to address "the status of women" does not bode well for free speech and transparency in related issues or programming.

Infrastructure

Djibouti is a poor country. Its population is largely concentrated in urban centers. The largest city and most densely populated area is Djibouti Ville with a population over 600,000. No other city in the country has a population over 50,000.[68] More than three quarters of Djiboutians live in cities and towns, and the rural population subsists primarily by nomadic herding. The country is prone to droughts and floods, and has few natural resources. Scarce rainfall and less than 4 per cent arable land means that

60 Djibouti Telecom (DT) is the sole provider of telecommunications services and utilizes mostly a microwave radio relay network. Central Intelligence Agency, "CIA World Factbook: Djibouti."
61 About 4 per 100 fixed-line teledensity and nearly 44 per 100 mobile-cellular. Central Intelligence Agency, "CIA World Factbook: Djibouti."
62 Central Intelligence Agency, "CIA World Factbook: Djibouti."
63 Reporters Without Borders, "Djibouti Index 2022."
64 Bertelsmann Stiftung, "2022 Djibouti Country Report," 10.
65 Ibid 18.
66 Reporters Without Borders, "Djibouti Index 2022."
67 Ibid.
68 Central Intelligence Agency, "CIA World Factbook: Djibouti."

most food must be imported, and limits farming to small quantities of fruit and vegetable production.[69] Djibouti's ports handle 95 per cent of Ethiopia's trade and service transshipments.[70]

Healthcare is limited by poor infrastructure, shortages of equipment and supplies, and a lack of qualified personnel, in addition to poor awareness of public health considerations and access to general education about and access to sanitation resources. These limitations are further exacerbated outside of urban centers. However, more than a third of health care recipients are migrants because healthcare in Djibouti is still better than those available in their neighboring home countries of Eritrea, Somalia, Syria, Yemen, and, more recently, Ethiopia.[71] This high demand, low supply profile of healthcare availability increases the risk of childbearing, sexual trauma, and FGM. Furthermore, poor infrastructure interferes with the collection and transfer of data pertaining to healthcare interventions and outcomes in general, let alone tracking disaggregated gender data as called for by the WPS agenda.

Physical Environment

Djibouti is a crossroads of migration to and from the Gulf countries and North Africa given its strategic location along the Gulf of Aden and the Red Sea. It hosts large numbers of refugees from Somalia and Yemen, as well as asylum seekers from Ethiopia.[72] A reported 100,000 people, mainly Ethiopians and some Somalis, travelled through Djibouti annually to attempt a dangerous sea crossing to Yemen, usually from the port of Obock.[73] However, this flow reversed from 2015 onwards as Yemenis began fleeing to Djibouti, with almost 20,000 arriving by August 2017. Most Yemeni refugees remain unregistered and choose to stay in the environs of Djibouti Ville rather than seeking asylum at a refugee camps.[74] Djibouti has hosted refugees and asylum seekers at camps for twenty years, despite a widespread lack of potable water, consistent food shortages, and rampant

69 Ibid.
70 Demissie, "Djibouti Looks to Ethiopia to Gauge its Economic Future."
71 Central Intelligence Agency, "CIA World Factbook: Djibouti."
72 Green and Snyder, *Stuck in Limbo.*
73 Central Intelligence Agency, "CIA World Factbook: Djibouti."
74 Ongoing challenges both inside camps and in the broader environment include inadequate supplies of potable water; water pollution; limited arable land; deforestation (with forests threatened by agriculture and the use of wood for fuel); and desertification; endangered species. Central Intelligence Agency, "CIA World Factbook: Djibouti."

unemployment.[75] As people are displaced into internally displaced persons (IDP) or refugee camps, rates of sexual violence are shown to rise.[76]

The 2020 Department of State country report on terrorism in Djibouti assessed that most travelers entering Djibouti do so by land at one of four border points.[77] The Djibouti National Police control border checkpoints (with support from Gendarmerie patrols between border posts), and the country's armed forces have responsibility for patrolling the country's borders in remote locations. Djiboutian law enforcement agencies coordinate their counterterrorism functions and information sharing internally.

Climate change is recognized as a driver of instability in the region. Climate change–related weather extremes and water scarcity in most parts of the region antagonize existing socioeconomic challenges. As the United Nations Development Program (UNDP) recognized, failing to adequately protect citizens from hardships and disasters related to climate change can "trigger frustration, lead to tensions between different ethnic and religious groups within countries and to political radicalization."[78] Desertification, drought, food insecurity, land degradation, and water scarcity are just some of the drivers that governments and other regional and international stakeholders must prepare themselves to address.[79]

Flash flooding has increasingly destroyed roads and other public infrastructure in recent years, and prolonged droughts are affecting pastoralist livelihoods and rural resilience.[80] Women and other marginalized groups are less able to survive and recover from climate shocks[81] because of higher levels of vulnerability and lower resiliency, and it has been recorded that women are 14 times more likely to die than men are in natural disasters.[82] Indeed, research shows that climate change is creating uneven burdens on women in addition to the unique vulnerabilities they face in conflict environments.[83]

---

75 Central Intelligence Agency, "CIA World Factbook: Djibouti."
76 UN Populations Fund, "Case Study: Somalia."
77 U.S. Department of State, "Country Reports on Terrorism: Djibouti."
78 United Nations Development Programme, "Addressing Radicalization and Violent Extremism through Climate Action."
79 Strong Cities Network, *The Role of Cities in Preventing and Countering Violent Extremism in East and West Africa*.
80 Bertelsmann Stiftung, "2022 Djibouti Country Report."
81 Chiu and Smith, "What Does Climate Change Have to do with Women, Peace, and Security? A Lot."
82 International Union for Conservation of Nature, "Disaster and Gender Statistics."
83 Chiu and Smith, "What Does Climate Change Have to do with Women, Peace, and Security? A Lot."

Time

The element of time would be scoped according to the operational profile of the entity producing the PMESII-PT application. In the context of this chapter, this dimension might track the emergence or evolution of legal and political rights for women, as well as changes in participation and representation across political and civil society.

While the above PMESII-PT product has taken gender considerations into account and is more intentionally sensitive to gender considerations than would generally be expected. Pursuing a more nuanced understanding of disproportionate circumstances faced by different population segments should be moved forward as a standard operating procedure and best practice in understanding operational environments. The above content is illustrative of how gender considerations can be integrated into existing tools.

Conclusion

While Djibouti maintains that all citizens are equal before the law without distinction to race, origin, sex, or religion, it is evident that gender equality is not a reality.[84] Having a NAP does not guarantee it. To date, states seem to have largely addressed the WPS agenda as a matter for Foreign Affairs or departments that have a focus on gender or women.[85] Both of these placements are potentially problematic, as there is inherent political significance of who is the lead actor named in a NAP. An outward orientation that positions the WPS as a matter of foreign policy, perhaps aimed at a conflict-affected area outside of the territory of the authoring state, may have a hypocritical 'othering' tone. While the positioning of WPS in gender or women-focused agencies, on the other hand, reinforces a demarcation between women's issues and peace and security, thus perpetuating the very framing that UNSCR 1325 is aimed at dismantling.[86]

While NAPs strive to address these issues at the strategic level, the inclusion of gender perspectives in PMESII-PT type evaluations of

84 Bertelsmann Stiftung, "2022 Djibouti Country Report," 26.
85 Hamilton, Naam, and Shepherd, *Twenty Years of Women, Peace and Security National Action Plans: Analysis and Lessons Learned*, 6.
86 Lee-Koo, "Implementing Australia's National Action Plan on United Nations Security Council resolution 1325."

the operating environment and the conduct of gender analysis allow for complementary responses in the nearer term at the operational and tactical levels. The explicit conduct of gender analyses should be accepted as best practice to sufficiently contextualize the disproportionate impacts of complex environments on women and other marginalized groups. Furthermore, national women's machineries often lack the sustained resources and political status in many states to attract genuine political interest and reliable support.[87] Indeed, cases such as the civil society and political mobilization seen in Liberia are the exception, rather than the norm.

WPS interventions cannot be stove-piped additions to programming if they are to be meaningful and durable. Rather, they must be integrated across whole-of-government planning and strive to educate their audiences about the importance of addressing the differing strengths and vulnerabilities of all population segments. Incorporating a gender analysis as a foundational component of environmental analysis will allow both host countries and partners to develop an expanded understanding of the operational environment, and to pursue opportunities and mitigate risks in turn.

References

Acheson, Ray. *Remote Warfare and Sexual Violence in Djibouti*, https://wilpf.org/wp-content/uploads/2017/09/2017_RemoteWarfareAndSexualViolenceInDjibouti.pdf.

Bertelsmann Stiftung. "2022 Djibouti Country Report." BTI Transformation Index. https://bti-project.org/en/reports/country-report/DJI.

Central Intelligence Agency. "CIA World Factbook: Djibouti." Updated October 4, 2022. https://www.cia.gov/the-world-factbook/countries/djibouti/.

Chiu, Clara and Jessica Smith. "What Does Climate Change Have to do with Women, Peace, and Security? A Lot." Women and Crisis. March 2, 2022. https://www.womenandcrisis.com/2022/03/what-does-climate-change-have-to-do.html.

Demissie, Selam Tadesse. "Djibouti Looks to Ethiopia to Gauge its Economic Future." Institute for Security Studies. May 3, 2021. https://issafrica.org/iss-today/djibouti-looks-to-ethiopia-to-gauge-its-economic-future.

Fleser, Cori. "Beyond Munitions: A Gender Analysis for Ukrainian Security Assistance." Atlantic Council. August 15, 2022. https://www.atlanticcouncil.org/in-depth-research-reports/issue-brief/beyond-munitions-a-gender-analysis-for-ukrainian-security-assistance/.

87 Swaine, "Assessing the Potential of National Action Plans to Advance Implementation of United Nations Security Council Resolution 1325," 417.

Freedom House. "Freedom in the World: Djibouti Country Report." 2022. https://freedomhouse.org/country/djibouti.

Fund for Peace. "Fragile States Index: Country Dashboard Djibouti." https://fragilestatesindex.org/country-data/.

Government of Canada. "Gender Analysis." April 23, 2019. https://www.international.gc.ca/world-monde/funding-financement/gender_analysis-analyse_comparative.aspx?lang=eng.

Green, Shannon N. and Julie N. Snyder. *Stuck in Limbo: Refugees, Migrants, and the Food Insecure in Djibouti.* Center for Strategic & International Studies. April 2017. https://csis-website-prod.s3.amazonaws.com/s3fs-public/publication/170418_Green_StuckInLimbo_Web.pdf.

Hamilton, Caitlin and Laura J. Shepherd. "WPS National Action Plans." LSE Centre for Women, Peace and Security. 2019. https://www.wpsnaps.org/.

Hamilton, Caitlin, Nylbeny Naam, and Laura J. Shepherd. *Twenty Years of Women, Peace and Security National Action Plans: Analysis and Lessons Learned.* (Sydney: University of Sydney, 2020). https://www.wpsnaps.org/app/uploads/2020/03/Twenty-Years-of-Women-Peace-and-Security-National-Action-Plans_Report_Final_Web.pdf.

International Monetary Fund. *Djibouti: Selected Issues.* Country Report 16/249. July 2016. https://www.imf.org/external/pubs/ft/scr/2016/cr16249.pdf.

International Union for Conservation of Nature. "Disaster and Gender Statistics." https://www.unisdr.org/files/48152_disasterandgenderstatistics.pdf.

Jacevic, Mirsad Miki. "WPS, States, and the National Action Plans." In *The Oxford Handbook of Women, Peace, and Security,* edited by Sara E Davies and Jacqui True, 273-90. Oxford: Oxford University Press, 2019.

Lee-Koo, Katrina. "Implementing Australia's National Action Plan on United Nations Security Council Resolution 1325." *Australian Journal of International Affairs,* 68, no. 3 (May 27, 2013): 300-13. DOI:10.1080/10357718.2014.901296.

Ministère de la Femme et de la Famille. *Plan d'Action National pour la Mise en œuvre de la Résolution 1325 (2000) et des Resolutions Connexes du Conseil de Sécurité des Nations Unies sur la Femme et la Paix.* November 2017. http://1325naps.peacewomen.org/wp-content/uploads/2021/07/Djibouti-2017.pdf.

Peacekeeping & Stability Operations Institute. "PKSOI WPS Info Sheet." U.S. Army War College. 2019.

Reese, Madeleine. "Investigating Foreign Military Bases and Sexual Violence in Djibouti." Huffington Post. September 27, 2017. https://www.huffpost.com/entry/investigating-foreign-military-bases-and-sexual-violence_b_59cb5c4de4b028e6bb0a66df.

Reporters Without Borders. "Djibouti Index 2022." https://rsf.org/en/country/djibouti.

Sahan. *The Somali Wire.* No. 269. November 15, 2021. Delivered electronically.

Strong Cities Network. *The Role of Cities in Preventing and Countering Violent Extremism in East and West Africa: An Assessment Report for the Strong Cities Network Prepared by the Global Center on Cooperative Security.* 2017. https://strongcitiesnetwork.org/en/wp-content/uploads/2017/10/SCN-Assessment-Report_East-and-West-Africa_June17.pdf.

Swaine, Aisling. "Assessing the Potential of National Action Plans to Advance Implementation of United Nations Security Council Resolution 1325." *Yearbook of International Humanitarian Law,* 12 (December 2009): 403-33. DOI:10.1017/S1389135909000142.

Transparency International. "Corruption Perceptions Index." 2021. https://www.transparency.org/en/cpi/2021/index/dji.

U.S. Agency for International Development. "Gender Equality and Women's Empowerment." Updated September 28, 2020. https://www.usaid.gov/djibouti/our-work/gender-equality-and-womens-empowerment.

U.S. Department of State. "Country Reports on Terrorism: Djibouti." https://www.state.gov/reports/country-reports-on-terrorism-2020/djibouti/.

U.S. Department of State. *Djibouti 2020 Human Rights Report.* March 30, 2021. https://www.state.gov/reports/2020-country-reports-on-human-rights-practices/djibouti/.

UN Migration. "Djibouti." https://www.iom.int/countries/djibouti.

UN Populations Fund. *Case Study: Somalia. Funding for Gender Equality and the Empowerment of Women and Girls in Humanitarian Programming.* June 2020. https://www.unfpa.org/sites/default/files/resource-pdf/GEEWG_UN_Women_Somalia_new.pdf.

UN Security Council. Resolution 2349, Peace and Security in Africa – Lake Chad Basin. S/RES/2349. (March 31, 2017). https://documents-dds-ny.un.org/doc/UNDOC/GEN/N17/086/16/PDF/N1708616.pdf?OpenElement.

UN Women. "Djibouti." https://data.unwomen.org/country/djibouti.

United Nations Development Programme. "Addressing Radicalization and Violent Extremism through Climate Action." November 14, 2016.

World Bank. "Djibouti's Economic Update – October 2021." October 7, 2021. https://www.worldbank.org/en/country/djibouti/publication/economic-update-october-2021.

7

MAKING SENSE OF ARMED CONFLICT

Gender Inequality and Climate Change in Military Operations

Jody M. Prescott and Robin Lovell[1]

All military operations require effective analysis of their mission areas, but many military organisations struggle in civilian-centric operations to accurately capture the complexity of cultural and social systems of host nations in actionable ways.[2] This is particularly true when seeking to analyse gender. In her influential work on gender analysis almost 30 years ago, Moser noted that

> many of those committed to integrating gender into their work at policy, programme and project levels still lack the necessary planning principles and methodological tools ... planners require simplified tools which allow them to feed the particular complexities of specific contexts into the planning process.[3]

The imperative to distil the actionable out of the incomprehensible by making some clarifying assumptions has not changed for military organisations, but in the case of gender it is obstructed by military doctrine that continues to be gender-blind, even if its proponents might style it

---

1 The authors gratefully acknowledge the outstanding work of Cameron Chambers, Emma Miner, Cyrus Oswald, and Emily Smith, students at the University of Vermont, done in conjunction with the research of this chapter. Their work is highlighted in the last section, which describes the details of the proposed model.

2 Jody M. Prescott, "Lawfare: Softwiring the Network," *The Three Swords* 32 (July 2017) 11-14.

3 Caroline O. N. Moser, *Gender Planning and Development: Theory, Practice and Training* (1st ed.) (London: Routledge, 1993), 10.

as gender neutral.[4] Analysis dealing with climate change is similarly hampered, because there is little evidence of the strategic recognition of its impacts trickling down to the operational level.[5] How then in an operational area marked by underlying gender inequality can deployed units meaningfully track whether women and girls are disproportionately experiencing the combined effects of war and climate change?

The international community has long recognized that armed conflict has different impacts on women, girls, men, and boys. The same holds true for climate change and its proxies, such as extreme weather events and food scarcity. As the effects of climate change continue to impact the global population, several researchers and writers[6] have expressed the concern that the simultaneous occurrence of armed conflict and climate change could be having a compounding effect on women and girls, particularly in regions of the world marked by gender inequality.[7] As one researcher has noted, however, those studying these relationships face a dilemma:

> What has been especially challenging has been to do so without, on the one hand, generalising to the point where one remains at the level of banal platitudes and, on the other, drowning in the complexity of local specificities.[8]

This chapter argues that this problem could be addressed by taking a socio-ecological systems (SES) approach to create a simplified tool for use by military organisations and others to better identify geographic areas at the highest risk of any compounding effects of armed conflict, climate change, and gender inequality. An SES is a "system that includes societal

4 Jody M. Prescott, "Gender Blindness in US Doctrine," *Parameters* 50, no. 4 (Winter 2020) 23-26.

5 Jody M. Prescott, *Armed Conflict, Women and Climate Change* (London: Routledge, 2019), 62-68.

6 UN Security Council Resolution 1325, On Women, Peace and Security, UN DOC. S/RES/1325 (31 October 2000); Judith Gardam and Michelle Jarvis, *Women, Armed Conflict and International Law* (The Hauge: Kluwer International, 2001), 20; Sahana Dharmapuri, "Just Add Women and Stir?" *Parameters* 41, no. 1 (2011) 64; Ann Jones, *War is Not Over when It is Over—Women Speak Out from the Ruins of War* (New York: Macmillan, 2010), 58, 161-62.

7 *See* Jessica M. Smith, Lauren Olosky, and Jennifer Grosman Fernández, "The Climate-Gender-Conflict Nexus – Amplifying women's contributions at the grassroots," Georgetown Institute for Women, Peace and Security, (2021), 5-14 (report), https://giwps.georgetown.edu/wp-content/uploads/2021/01/The-Climate-Gender-Conflict-Nexus.pdf; Mayesha Alam, "A cross-cutting agenda: Gender, climate change and conflict," ECDPM *Great Insights Magazine* 8, no. 4 (Autumn 2019): 27-30.

8 Henri Myrrtinen, "Drowning in complexity? Preliminary Findings on Gender, Peacebuilding and Climate Change in Honduras," in *Addressing Global Environmental Challenges from a Peace Ecology Perspective*, The Anthropocene: Politik—Economics—Society—Science 4 (Cham, Switzerland: Springer International Publishing, 2016), 98.

(human) and ecological (biophysical) subsystems in mutual interaction."[9] Importantly, this model does not attempt to identify and explain all the possible interactions and their effects. Instead, for purposes of clarity, an SES assumes that the impacts of armed conflict, climate change, and existing gender inequality exacerbate each other, and that they are additive to each other to some degree.

By identifying areas where the risk of these negative impacts is greatest, the model could be used prior to deployment to determine whether additional training for personnel on gender factors in the mission area is needed. This would further the prevention strand of the Women, Peace and Security (WPS) agenda as well as contribute to operational effectiveness. It could also be used to recommend initial courses of action even in the absence of definitive gender-disaggregated data and analysis. Mindful of this absence, however, it could also be useful in guiding the investment of focused data collection and analysis in these high-risk areas to better understand what was actually happening on the ground.

The chapter will proceed as follows. The next section explores the relationships between climate change, gender inequality, and armed conflict, and in particular the impacts of these relationships upon perhaps the most vulnerable humans – children. Because of children's sensitivity, metrics concerning impacts on them should help register fine but important changes in the operational area. A discussion of taking an SES approach to gender analysis then follows, contrasting its utility with NATO's military gender analysis methodology as found in the Allied Command Operations' *Gender Functional Planning Guide*.[10] Using an SES approach, the last section then justifies the format and organization of the model, including proposed criteria for all layers, and describing in detail the intervals for the gender inequality layer as a working example.

In sum, the proposed model is this: a visual representation of risk formed by the sequential stacking of three colour-density coded overlays atop one another: first Gender Inequality; then Armed Conflict; and lastly Climate Change. Each overlay would be based on data collected against the four or five most relevant criteria, with each criterion breaking down into

9 Gilberto C. Gallopín, "Linkages between vulnerability, resilience, and adaptive capacity," *Global Environmental Change* 16, no. 3 (2006): 293-303, 294.

10 NATO, Allied Command Operations, *Gender Functional Planning Guide* (Brussels: Supreme Allied Headquarters Europe, July 2015).

four or five intervals.[11] Again, the proposed model works off the assumption of the additive effects of climate change, conflict, and gender disparities, and is meant to help pre-operative planning for more effective outcomes.

The conclusion addresses some of the shortcomings with the proposed model and recommends steps for further study to test the model's usefulness. Specifically, whilst climate change may not be immediately threatening to an operation, it is worth reflecting on the secondary and compounding effects it may have on armed conflict and gender. From a utilitarian perspective, the proposed model could save operational analysts time and help them include more detail than is standard in a typical NATO-style gender analysis.

Armed Conflict, Gender Inequality, and Climate Change: Do the Relationships Even Matter?

Before exploring the armed conflict and climate change impacts that affect women, and especially children, and in particular exploring research that suggests that the two forces could compound each other in a gender differentiated way, this chapter first addresses whether it is worthwhile from an operational perspective to even start going down this path. Within the military operational context, the intersections of armed conflict, gender inequality, and climate change remain largely unexamined. Gender-blind military doctrine that does not promote the operational significance of collecting sex and gender-disaggregated data is just one factor.[12] The vastness and complexity of the human experience of gender that is found across the range of operational situations does not lend itself to easy data collection and analysis.

In this light, it is not surprising that a recent scientific study – by two highly respected climate change experts – that examined the linkages between armed conflict, climate change and population-level vulnerabilities to those two forces largely dismissed the significance of

11 See Jody M. Prescott, Robin Lovell, and Team Sandhurst (Cameron Chambers, Emma Miner, Cyrus Oswald, Emily Smith), "Gender inequality, armed conflict and climate change: why militaries can and should map compounded risk," International Committee of the Red Cross, *Humanitarian Law & Policy Blog* (29 September 2022), https://blogs.icrc.org/law-and-policy/2022/09/29/gender-inequality-armed-conflict-climate-change-militaries-map-compounded-risk/ (graphic depiction of stacked overlays).

12 See for example, Headquarters, Department of the Army (HQDA), *Field Manual 3-57, Civil Affairs Operations* (Washington, D.C.: HQDA, 2021).

gender in understanding these relationships. In this study, Buhaug and Von Uexkull identified societal and political factors as the two primary determinants of vulnerability to the combined effects of armed conflict and climate change in a conceptual model of the intersections between violence, "vulnerabilities", and climate change.[13] They identified a third category as "other generic determinants of vulnerability," which included gender inequality, rapid population growth, and biases in risk perception. They excluded these drivers of vulnerability from further analysis, because these factors were "less important in shaping conflict risk and they also tend to be less impacted by climatic shocks and armed conflict than socioeconomic and political conditions".[14]

There are three potential problems with this assessment from the perspective of determining whether there is operational significance to the potentially compounding effects of armed conflict and climate change in the presence of gender inequality. First, although our literature review shows that gender inequality does not significantly influence the risk of conflict, this analysis is more concerned with effectively dealing with the actual impacts of armed conflict and climate change in areas marked by gender inequality. In other words, the approach takes a snapshot of how existing vulnerabilities may influence operations, rather than predicting the onset or degree of conflict. Second, the socioeconomic and political factors Buhaug and Von Uexkull address at a population level as being the primary determinants have been shown to be affected by gender inequality within populations.[15] It is difficult to imagine that something that affects so many people negatively does not impact the analytical significance of it to a society. These factors, such as poverty, dramatically impact women's ability to cope with shocks. For example, there is a strong trend globally called the "feminization of poverty," in which growing income inequality impacts women disproportionately.[16] Within this trend, poverty does not simply refer to a woman's income, but also other wellbeing factors such as representation and participation in governance, systematic exclusion from access to and control of resources, literacy, and the ability to secure

13 Halvard Buhaug and Nina von Uexkull, "Vicious Circles: Violence, Vulnerability, and Climate Change," *Annual Review of Environment and Resources* 46 (2021): 549-50.

14 Buhaug and Von Uexkull, "Vicious Circles," 550-51.

15 Prescott, *Armed Conflict*, 15-54.

16 Sara S. McLanahan and Erin L. Kelly, "The feminization of poverty," in *Handbook of the Sociology of Gender* ed. Janet Saltzman Chafetz (Boston, MA: Springer, 2006); 127-145; Sakiko Fukuda-Parr, "What does feminization of poverty mean? It isn't just lack of income," *Feminist Economics* 5, no. 2 (1999): 99-103.

financial resources. Third, there are a significant number of studies in the scientific literature that show that armed conflict and climate do indeed have gender-differentiated impacts, and that women and girls in general experience different and more severe effects from both.[17]

Mindful of these potential problems, this chapter recognizes that working gender inequality into a conceptual model intended for broad applicability to the study of the relationships between armed conflict and climate change presents challenges. Although the more severe impacts of both upon women are well-catalogued, in any given mission area, not all these negative effects will occur at the same time and in the same place. They might even not occur at all. This variability places a premium on avoiding generalizations and tailoring analysis to the characteristics of particular regions.

The effects women experience from participating in armed conflict expand the impact of existing climate and gender interactions of armed conflict beyond civilians. It has been estimated that women made up between 20 and 40 per cent of fighters in the Colombian non-state armed group *Fuerzas Armadas Revolucionarias de Colombia* (FARC) and perhaps as much 40 per cent in the former Maoist People's Liberation Army in Nepal.[18] Further, the effects upon combatant women are highly variable. For example, women combatants in many non-state armed forces are at risk of becoming "bush wives" to the men fighters of their group and experiencing harsh social stigmas when they return to their families, especially if they bring children with them from their wartime relationships.[19] This does not seem to be an issue, however, for the women fighters of different Kurdish non-state armed groups, who are respected in their communities for their armed service and are not placed in positions where they find themselves sexually subservient to their male colleagues. This range of social drivers and outcomes suggests caution is necessary in applying generalities regarding gendered experiences in any specific analysis of an operational area.

However, simply because it might be difficult to apply a conceptual model to something does not mean it is without operational significance,

17 Prescott, *Armed Conflict*, 15-54. It is also important to note that one of the criticisms of scientific studies of the relationships between climate change and armed conflict is that they have tended to not factor in the role of gender. Roz Price, "Climate Change as a driver of conflict in Afghanistan and other Fragile and Conflict Affected States, *KD4 Helpdesk Report* (18 January 2019): 2, https://assets.publishing.service.gov.uk/media/5ca20ba940f0b625df8d85f1/527_Climate_change_as_a_driver_of_conflict_in_Afghanistan_and_other_FCAS.pdf.
18 Myrrtinen, "Drowning in complexity?" 99 n. 2.
19 Prescott, *Armed Conflict*, 21–24.

or that it cannot be effectively dealt with during prudent military planning. It is not necessary for commanders and planners to have definitive understanding of operational factors so long as they understand that these operational factors could be threats. In such cases, investments in simplified procedures and tools to provide sufficient leverage to work with such factors can be worthwhile. To develop the approach, there must be an adequate factual basis to define the threat to mission accomplishment, and the things that are being measured must be particularly telling. The impacts upon children are among the items that meet these criteria.

Children, Gender Inequality, Armed Conflict, and Climate Change

The literature regarding the impacts of armed conflict and climate change upon children is important not just because of the potential gendered effects upon them, but also because in many societies, women largely take on the responsibility for childcare. Children are likely to bear the brunt of many of the hardships caused by armed conflict and climate change. Their suffering will not occur in a vacuum; it will often be reflected in additional work and stress for their under-resourced mothers and other female caregivers, degrading their health and economic and educational opportunities.[20]

These burdens rarely directly shape the risk of conflict, but in civilian-centric missions, they undermine efforts to build the human security necessary for stability and conflict transformation. Factoring in gendered experiences is therefore crucial for the effective and efficient execution of such missions. Many studies have documented that investment in social and economic stability of women and girls has a significant, positive impact on the wellbeing of said society (life expectancy, literacy, income, etc.) compared to investment in men.[21] In other words, giving women resources, information, and assistance helps societal progression more than giving the same to men. It is in this context that we examine the impacts armed conflict has on children, physically and mentally.

20 Anil Patil, "Why caregiving needs to be 'de-feminized,'" White Swan Foundation (28 March 2018), https://www.whiteswanfoundation.org/caregiving/how-the-role-of-0caring-disproportionately-burdens-women.

21 Daraka Chhay, "Women's economic empowerment through microfinance in Cambodia," *Development in Practice* 21, no. 8 (2011): 1122-1137; Boudor Bzeouicha, "Microfinance and The Improvement of Social Welfare of Beneficiary Households: Evidence from Tunisia," *Biltürk Journal of Economics and Related Studies* 1, no. 3 (2019): 188-205.

The Impacts of Armed Conflict and Climate Change on Children

Fetuses are particularly sensitive to the impacts of toxins and diseases because of their rapid development *in utero*.[22] The negative impacts upon children whose mothers experienced armed conflict while they were *in utero* is well documented. Duque, for example, determined that Colombian children who had been exposed to terrorist attacks *in utero*, even as early as the first trimester of their mothers' pregnancies, were negatively affected with low birth weight and slower physical and cognitive development.[23] Although her study did not directly measure stress, she noted that the results she found were "consistent with stress being an important mechanism driving these effects."[24]

Duque's findings were largely confirmed by Guantai and Kijima in their study of Kenyan babies whose mothers experienced ethnic conflict in 1992. They found that not only were babies who were exposed to violence in the womb more likely to have lower birth weight, they were also likely to be smaller in size.[25] Importantly, their data did not find that these outcomes were mainly the result of educational differences between the mothers, poor nutrition during the fighting, or even availability of health care. They concluded, instead, that exposure to violence caused maternal stress and this was the main driver of these results.[26] Interestingly, Tsujimoto and Kijima's study on the *in utero* impacts of violence of the 1990-1994 conflict in northern Mali found no significant difference in baby birth weight but did confirm a significant difference in the babies' size.[27] This variability in findings serves as a reminder of the importance of not assuming for purposes of operational analysis that results valid in one area are necessarily valid in others.

Climatic conditions also impact fetuses. Kumar and his co-authors found that among rural Indian populations, children who were exposed to drought *in utero* were more likely to experience both stunting and being

22 Perry E. Sheffield and Phillip J. Landrigan, "Global Climate Change and Children's Health: Threats and Strategies for Prevention," *Environmental Health Perspectives* 119, no. 3 (March 2011): 291.

23 Valentina Duque, "Early-life conditions and child development: Evidence from a violent conflict," *SSM – Population Health* 3 (2017): 121-122.

24 Duque, "Early-life conditions," 131.

25 Fredah Guantai and Yoko Kijima, "Ethnic Violence and Birth Outcomes: Evidence From Exposure to the 1992 Conflict in Kenya," *Demography* 57 (2020): 423.

26 Guantai and Kijima, "Ethnic Violence and Birth Outcomes," 442.

27 Takahiro Tsujimoto and Yoko Kijima, "Effects of conflict on child health: Evidence from the 1990-1994 Northern Mali Conflict," *Health Economics* 29 (2020): 1456.

underweight and have a higher probability of not surviving to their first birthdays.[28] This was particularly the case for those fetuses exposed to the impacts of the drought during the first and third trimesters of their mothers' pregnancies, as a result of decreased agricultural production.[29] Other researchers have found that high temperatures can increase the risk of pre-term births.[30] Furthermore, extreme temperatures pose special risks for pregnant women and their fetuses because of the additional thermal stress posed by the changes in women's bodies as a result of pregnancy.[31] The impacts of extreme heat do not end at birth, however, because extreme heat during any trimester tends to result in lower birth weight, which is often associated with illness both during childhood and adulthood.[32]

Finally, climate change has resulted in greater exposure of populations to wildfires, which is associated with a greater risk of maternal illness,[33] greater risk of premature birth[34] and reduced birth weight.[35] There is also a strong association between exposure to wildfires in pregnancy and birth defects, "[w]ildfire exposure within 30 days before pregnancy and was associated with more than two times higher risk of foetal gastroschisis, and a 28 per cent higher risk of this if the exposure was in the first trimester."[36]

For children, armed conflict negatively impacts their mental health, resulting in increased instances of post-traumatic stress disorder and

28 Santosh Kumar, Ramona Molitor, and Sebastian Vollmer, "Drought and Early Child Health in Rural India," *Population and Development Review* 42, no. 1 (March 2016): 65.

29 Kumar *et al.*, "Drought and Early Child Health," 63.

30 Jegosothy, Edward, Deborah A. Randall, Jane B. Ford, Tanya A. Nipita, and Geoffrey G. Morgan, "Maternal factors and risk of spontaneous preterm birth due to high ambient temperatures in New South Wales, Australia," *Paediatric and Perinatal Epidemiology* 36, no. 1 (January 22): 4-12.

31 Joshua Graff Sivin and Jeffrey Shrader, "Temperature Extremes, Health, and Human Capital," *The Future of Children* 26, no. 1 (Spring 2016): 37.

32 Sivin and Shrader, "Temperature Extremes, Health, and Human Capital," 38.

33 Korin Miller, "How Pregnant women and Babies Can Stay Safe in Wildfire Smoke," *Whattoexpect.com* (17 September 2020), https://www.whattoexpect.com/news/family/wildfire-smoke-safety-pregnant-women-babies (citing the U.S. Centers for Disease Control).

34 Sam Heft-Neal, Anne Driscoll, Wei Yang, Gary Shaw, and Marshall Burke, "Associations between wildfire smoke exposure during pregnancy and risk of preterm birth in California," *Environmental Research* 203 (January 2022): 8.

35 Weeberb J. Requia, Heresh Amini, Matthew D. Adams, and Joel D. Schwarz, "Birth weight following pregnancy wildfire smoke exposure in more than 1.5 million newborns in Brazil: A nationwide case-control study," *The Lancet Regional Health, Americas* (14 March 2022), https://www.thelancet.com/journals/lanam/article/PIIS2667-193X(22)00046-1/fulltext.

36 Bo Young Park, Ian Boles, Samira Monavvari, Shivani Patel, *et al.*, "The association between wildfire exposure in pregnancy and foetal gastroschisis: A population-based cohort study," *Paediatric and Perinatal Epidemiology* 36, no. 1 (January 2022): 45-53.

depression.[37] This is particularly pronounced among those children who are forced to fight as child soldiers.[38] Similarly, children are particularly sensitive to the impacts of climate change, in part because of their smaller size and developing body systems, in part because of their reliance on their caregivers for survival.[39] Young children are particularly sensitive to extreme heat.[40] Increased heat can lead to babies gaining weight too fast, which increases their risk of being obese later in life.[41] Furthermore, high temperatures have been linked to premature births, which can result in life-long health effects and increased need for medical care.[42]

Perhaps one of the greatest risks to children is the increased likelihood of malnutrition as a result of either war or climate change. Akresh *et al.*, for example, found that Eritrean boys and girls who were born during the 1998-2000 conflict between Eritrea and Ethiopia experienced stunting, or low height for their age.[43] The longer children are exposed to the effects of armed conflict, the greater the degree of stunting they might suffer.[44] A number of these impacts are also gender-differentiated and generational. For example, women who suffered malnutrition as children are at increased risk of having babies who are born prematurely or underweight.[45]

---

37 Rochelle L. Frounfelker, Nargis Islam, Joseph Falcone, Jordan Farrar, Chekufa Ra, Cara M. Antonaccio, Ngozi Enelamah and Therese S. Betancourt, "Living through war: Mental health of children and youth in conflict-affected areas," *International Review of the Red Cross* 101, no. 911 (2019): 484.

38 Frounfelker *et al.*, "Living through war," 486.

39 Perry E. Sheffield and Phillip J. Landrigan, "Global Climate Change and Children's Health: Threats and Strategies for Prevention," Environmental Health Perspectives 119, no. 3 (March 2011): 291.

40 Joshua Graff Sivin and Jeffrey Shrader, "Temperature Extremes, Health, and Human Capital," *The Future of Children* 26, no. 1 (Spring 2016): 34-35.

41 Carlos Francisco Dionicio López, Neora Alterman, Ronit Calderdon-Margalit, Michael Hauzer, *et al.*, "Postnatal exposure to ambient temperature and rapid weight gain among infants delivered at term gestations: a population-based cohort study," *Paediatric and Perinatal Epidemiology*, 36, no. 1, (January 2022): 26-35.

42 Edward Jegasothy, Deborah A. Randall, Jane B. Ford, Tanya A. Nipita, and Geoffrey G. Morgan, "Maternal factors and risk of spontaneous preterm birth due to high ambient temperatures in New South Wales, Australia," *Paediatric and Perinatal Epidemiology* 36, no. 1 (January 2022): 4-12; Lara Cushing, Rachel Morello-Frosch, and Alan Hubbard, "Extreme heat and its association with social disparities in the risk of spontaneous preterm birth," *Paediatric and Perinatal Epidemiology* 36, no.1 (January 2022): 13-22.

43 Richard Akresh, Leonardo Lucchetti, Harsha Thirumurthy, "Wars and child Health: Evidence from the Eritrean-Ethiopian Conflict," *Journal of Development Economics* 99, no. 2 (November 2012): 336-337.

44 Tom Bundervoet, Philip Verwimp, and Richard Akresh, "Health and Civil war in Rural Burundi," *The Journal of Human Resources* 44, no. 2 (2009): 537.

45 United Nations Standing Committee on Nutrition, *Sixth report on the world nutrition situation: Progress in Nutrition* (2010): 11, 53, 64.

In their recent review of 90 different studies related to malnutrition in children, Brown and her co-authors focused on studies that addressed at least one of three internationally recognized malnutrition indicators: wasting (low weight for height), stunting, and underweight (low weight for age).[46] In these studies, climate conditions were frequently included in the analyses, and conflict conditions were included to a lesser degree. Brown *et al.* recognized that although the relevant literature indicated that climate change effects and conflict exacerbate each other, there were few studies that considered the joint effects of climate and conflict upon children.[47] In their analysis, they found that there were statistically strong associations between wasting and excessive rainfall, between stunting and extreme temperatures, and between underweight and drought.[48] Conflict, assessed using various metrics, was strongly associated as a risk factor for stunting.

They also found that "[a]mong the consistent predictors of child malnutrition are shocks due to variations in climate conditions (as measured with indicators of temperature, rainfall, and vegetation) and violent conflict."[49] Importantly, however, they identified three factors that could mitigate these impacts: the children's mothers being better educated, having a greater body mass index, and being more affluent.[50] From an operational analysis perspective, where both time and data are scarce, information on children's health then might serve as a decent proxy for assessing the degree to which women in a particular area are at potentially greater risk to any compounding impacts of climate change and armed conflict in the near term.

The health risks to children's well-being caused by armed conflict and climate change are reflected in education. Ordinarily, these factors are found to decrease educational opportunities for girls.[51] However, two examples in this area show how important it is to avoid generalities in gender analysis. In Valente's study of children's educational attainment in Nepal during the Maoist insurgency, it appears that armed conflict led to an increase

46 Molly E. Brown, David Backer, Trey Billing, Peter White, Kathryn Grace, Shannon Doocy, and Paul Huth, "Empirical studies of factors associated with child nutrition: highlighting the evidence about climate and conflict shocks," *Food Security* 12, no. 5 (2020): 1242. Wasting occurs when a child with ordinarily good nutrition suffers a rapid and substantial weight loss. Stunting, on the other hand, suggests a long period of malnutrition, and its effects might be irreversible. A child who is underweight could be experiencing wasting, stunting, or perhaps both.

47 Brown *et al.*, "Empirical Studies of factors," 1248.

48 Ibid., 1246.

49 Ibid., 1241.

50 Ibid., 1246.

51 Prescott, *Armed Conflict*, 19, 40-41, 48,

in educational attainment for both girls and boys, but more so for girls.[52] Valente proposes this was due in part to the insurgents' insistence in areas they controlled on gender equality and education, consistent with their ideology.[53] Further, it must be noted that the insurgent forces included significant numbers of female personnel.[54]

Conversely, when the insurgents abducted school children in their mass indoctrination operations, girls' educational attainment decreased in those areas.[55] Valente posits that perhaps it was because parents were concerned about their daughters' chances for favourable marriages if they had been raped or believed to have been raped, and therefore kept them away from school to lessen the chances of abduction.[56] Conceivably, depending on the conflict, this could be happening simultaneously in an area of operations. A study of educational attainment among children in Timor-Leste in the wake of the 1999 violence also found that girls accomplished more than their brothers. The study suggests that, as a result of conflict, parents were steering sons towards participation in the workforce whereas daughters were able to take advantage of schools put in place by the transitional international authority.[57]

The Potential Compounding Effects of Armed Conflict and Climate Change

There are relatively few studies of the combined effects of armed conflict and climate change on at-risk populations, but for purposes of the proposed model they are worthwhile to review. In their study of the impacts of political violence and cyclical droughts on children in the Indian state of Andhra Pradesh, Tranchant *et al.* find the health of children in villages that experienced only the recurring droughts does not seem to have significantly suffered, as measured by height-for-age scores (a measure of long-term malnutrition) and to a lesser degree weight-for-age scores (an

52 Christine Valente, "Education and Civil Conflict in Nepal," *The World Bank Economic Review* 28, no. 2 (2014): 354.
53 Ibid., 355, 379.
54 Ibid., 380.
55 Ibid., 379.
56 Ibid., 378.
57 Mayra Buvinić, Monica Das Gupta and Olga N. Shemyakina, "Armed Conflict, Gender and Schooling," *The World Bank Economic Review* 28, no. 2 (2014), 316-317.

indicator of short-term malnutrition).[58] In those villages that experienced political violence because of the long-running insurgency and drought, however, children's nutrition did suffer.[59] The authors concluded that in this case study there was no evidence that the children's exposure to conflict alone had any significant impacts on their nutrition, and that data from an eight-month ceasefire period in 2004 showed that the climatic shocks on child nutrition were completely offset by the ceasefire. They also concluded that the political violence had made it harder for those villages to cope with the economic impacts of the drought and because of the restricted access to public goods and services.[60] Although the data on the children was not disaggregated by sex, it appears that political violence at the levels experienced had a measurable additive effect on at-risk populations in the presence of climate shock.

In another study, Akresh *et al.* looked at conflict and crop failure in Rwanda in 1990 and the effects on children under the age of five. Families in the area impacted by the crop failure appear to have chosen to shield their sons from food scarcity, but not their daughters.[61] Digging deeper into the height-for-age data showed that it was really the girls who belonged to poorer families who suffered, whilst wealthier families, measured by livestock holdings, provided for all their children.[62] As a result of the fighting that occurred in other areas of Rwanda, all children were negatively impacted health-wise.[63] The study's authors suggest that this is because the fighting was unexpected and severe, and deprived parents of their food supplies through theft and displacement irrespective of their socio-economic standing.[64] Further displacement likely made children more susceptible to disease.[65] Assuming this is correct, it also suggests that there is a gradient of compounding impacts of armed conflict and climate change upon at-risk populations – at a certain point, either could be so severe as

58 Jean-Pierre Tranchant, Patricia Justino, and Cathérine Müller, "Political Violence, Drought and Child Malnutrition: Empirical Evidence from Andhra Pradesh, India," HiCN Working Paper No. 173 (April 2014) 2–3, 13–14 (Households in Conflict Network, The Institute of Development Studies (University of Sussex), https://www.hicn.org/wp-content/uploads/sites/10/2012/06/HiCN-WP-173.pdf.
59 Ibid., 15.
60 Ibid., 21-23.
61 Richard Akresh, Philip Verwimp and Tom Bundervoet, "Civil War, Crop Failure, and Child Stunting in Rwanda," *Economic Development and Cultural Change* 59, no. 4 (2011): 777-78.
62 Ibid., 778, 787, 795.
63 Ibid., 790.
64 Ibid., 805.
65 Ibid.

to effectively cancel out any additive effect from the other. This edge must also be factored into any operational analysis of potential compounding effects upon the people in an area of operations.

Summary

The body of research on the gender-differentiated impacts of armed conflict establishes that in general each affects women and girls differently and more severely than men and boys. Within this more granular picture of the gender-differentiated interplays within these two dyads the generalities noted earlier often resolve themselves into pictures of greater complexity. It requires little imagination to see how this complexity could be confusing for members of a planning staff who were not well versed in gender research and application. Further, as demonstrated by a frequently used military gender analysis approach, NATO's gender functional planning guidance, it is not clear that there are existing tools well-suited to help them navigate this complexity. That is why an SES analytical approach could be a more effective and efficient instead.

NATO's Military Gender Analysis Method vs. a Socio-Ecological Systems Approach

Existing military gender analysis methodologies largely fail to bridge the gap between the existence of regional and national instances of gender inequality and the provision of actionable intelligence at the operational and tactical levels to systematically do something about them. The relationships between pairs of factors are often neither linear nor direct, and in the pairing of climate change and armed conflict alone, causality is heavily mediated by area-specific social and environmental factors.[66] As an example of a current military gender analysis method, this section explains why NATO's approach to analysing gender will not likely suffice to address these intersections, and why an SES approach could.

66 Katharine J. Mach *et al.*, "Climate as a risk factor for armed conflict," *Nature* 571 (2019): 193-95.

NATO Gender Analysis

NATO's approach leans heavily toward role analysis and relies on relational analysis to a much lesser degree. To understand NATO gender analysis, the Bi-Strategic Command Directive refers the reader to Annex E of the Gender Functional Planning Guide produced by the operational strategic command, Allied Command Operations.[67] The Gender Analysis Tool in this annex relies primarily on the PMESII model of operational analysis that regular staff sections in a headquarters would use as required by the overarching Comprehensive Operational Planning Directive.[68] The PMESII model analyses six 'domains' of an operational area, political, military, economic, social, infrastructure, and information.[69] As set out in the annex of the Gender Functional Planning Guidance, the examples of the gender factors considered in these domains are largely quantifiable, for instance in the Social Domain, the "[n]umber of men, women, boys and girls in the population."[70] The annex does describe four 'dimensions' that can be considered in relation to the domains, including the Relations Dimension that addresses relationships between different actors in the domains, but its definition provides neither gender-specific examples of what it means or how to account for these relationships' significance.[71]

Thus, standard NATO gender analysis is not well-suited to analysing the complex relationships between social and environmental factors that characterize the intersections between armed conflict, gender inequality, and climate change and environmental degradation. The Bi-Strategic Command Directive that defines gender analysis states that it "requires the systematic gathering and examination of information on gender differences and on social relations between men and women in order to identify inequities based on gender."[72] On its face, it appears that NATO's preferred mode of gender analysis is therefore a hybrid, one that would couple the roles-heavy approach set out in the Harvard Analytical Framework with

---

67 NATO, Allied Command Operations and Allied Command Transformation, *Bi-SC Directive 040-001, Integrating UNSCR 1325 and Gender Perspective into the NATO Command Structure* (Brussels and Norfolk, VA: Supreme Allied Headquarters Europe and Supreme Allied Command Transformation, 17 October 2017), 5 n.8.

68 NATO, Allied Command Operations *Comprehensive Operations Planning Directive Interim Version 2.0* (4 October 2013).

69 NATO, *Gender Functional Planning Guide*, E-2.

70 Ibid., E-3.

71 Ibid., E-3.

72 NATO, *Bi-SC Directive 040-001*, 5.

the Social Relations Approach that looks at a community primarily in terms of how its members of different genders interact with one another.[73]

This reliance on the roles-heavy PMESII approach for gender analysis also appears to be reflected in the training on developing gender analyses given to students at the Nordic Centre for Gender in Military Operations, which is NATO's department head for gender education and training.[74] This approach is influential, and it does have the benefit of following the same analytical model other staff sections would use to analyse their slices of the operational environment. However, when it is applied to gender in this way, it creates silos of gender information, and within these silos the information is stratified. This is not an effective way to analyse intersectional relationships. For these reasons, an SES approach is better suited to explaining the relationships between gender inequality, armed conflict and climate change. Although not traditionally used in military analysis, its strength lies in working with intersections, particularly those involving the natural world.

The SES Approach

The origins of SES research stem back to Holling's landscape ecology principles, which viewed ecosystems as more or less resilient depending on their ability to respond to perturbations. Holling defined resilience as "a measure of the persistence of systems and of their ability to absorb change and disturbance and still maintain the same relationships between populations or state variables."[75] Holling and Meffe later distinguished between two types of resilience, which represent the two epistemological approaches to ecosystem states over time.[76]

The first is *equilibrium resilience* in which speed of return to a steady-state and resistance to change from an equilibrium are the measure of resilience. They traced the obsession with equilibrium back to deductive mathematical theory. Recently, this approach is more often equated with the ecological concept of resistance. Within the equilibrium camp, resilience

73 Candida March, Ines Smyth, and Maitrayee Mukhopadhyay, *A Guide to Gender-Analysis Frameworks* (Oxford: Oxfam, 1999), 22–25.
74 Annika Berg, "Gender analysis—Planning Phase 2," Nordic Centre for Gender in Military Operations, Powerpoint Presentation (undated, but prior to March 2020).
75 C. S. Holling, "Resilience and stability of ecological systems," *Annual Review of Ecology and Systematics*, 4 (1973): 14.
76 C. S. Holling and Gary K. Meffe, "Command and Control and the Pathology of Natural Resource Management," *Conservation Biology* 10, no. 2 (1996): 328-337.

is conceptualized as "the ability of the system to maintain its identity in the face of internal change and external shocks and disturbances."[77]

The second is *ecosystem resilience*, which measures the amount a system can absorb before the fundamental processes, variables, or system behavior change. This tradition stems from landscape studies, inductive reasoning, and a broader understanding of scales of interaction. This concept of non-equilibrium or multi-stable systems revolutionized both ecological and SES studies, and it is the spirit of the proposed model. In other words, there are multiple paths towards a resilient SES, including military objectives and outcomes, but that accounting for the complexity civilian interactions bring is essential for successful peacekeeping missions.

Summary

Finding the best way to meaningfully incorporate human security concerns into military activities and operations is no easy task. Both gender and climate change have struggled to find their footing in military doctrine, and there is little recognition that the intersections between armed conflict, gender inequality and climate change present specific and largely unmeasured security risks in civilian-centric operations. The high degree to which gender concerns[78] and climate variability[79] have become controversial political issues over the last decade has likely been partly responsible for that, as has the re-emergence of strategic competition between great powers with its focus on expensive technology and force-on-force engagements.[80] Thus, a staff product that seeks to operationalise these concerns rather than just discuss them must demonstrate utility. It must be both simple to apply and to understand. Its methodology must be transparent and objective, and it must generate tangible, actionable recommendations for courses of action and resource allocations. That is why a model that takes an SES approach, which does not dumb down the

---

77 Cumming, G., G. Barnes, Stephen G. Perz, M. Schmink, Kathryn E. Sieving, J. Southworth, Michael W. Binford, R. D. Holt, C. Stickler, and T. V. Holt, "An Exploratory Framework for the Empirical Measurement of Resilience," *Ecosystems* 8 (2005): 975.

78 Judith Butler, "Why is the idea of 'gender' provoking backlash the world over?" *The Guardian*, 23 October 2021, https://www.theguardian.com/us-news/commentisfree/2021/oct/23/judith-butler-gender-ideology-backlash.

79 ElaineKamarck,"Thechallengingpoliticsofclimatechange,"BrookingsInstitute,23September 2019, https://www.brookings.edu/research/the-challenging-politics-of-climate-change/.

80 Ronald O'Rourke, "Renewed Great Power Competition: Implications for Defense—Issues for Congress," *Congressional Research Service* (21 December 2021): 1–15, R43838.pdf (fas.org).

complexities, and instead presents an easily understood picture of where the risks of the complexities' interactions pose threats to human security could be valuable for military operational effectiveness.

The Overlay Model

An SES Approach

An SES approach is well-suited to crafting a simplified gender analysis. The tool could support recommendations for pre-deployment training on gender; more refined data collection and analysis plans and intelligence requests; and informed recommendations for the deployment of personnel, units and, equipment in theatre. These planning benefits could mitigate the potential risks posed to a mission because of the potential compounding effects of armed conflict and climate change on women and girls.

In a particularly insightful approach to measuring SES outcomes, Ostrom proposes a nested, multitier framework upon which simplified set of overlays could be based. The framework attempts to account for systems (fisheries, dictatorships, and capitalism), the resource units generated by those systems (fish, corruption, and per capita income), the users (civilian and military personnel, in this case), and the governance structures who both affect and are influenced by the interactions in the system.[81] The simplified version of this approach is set out below (Figure 7.1) across a range of criteria to populate the overlays as an alternative to NATO-style PMESII analysis. This example focuses on fish, but the analysis could begin with any component of the framework, such as the fish resources units (RU) and work backwards to fill out the interactive effects of the system components. The idea is to think holistically about how all the components of a system interact.

The Model

The model starts with a base layer of a map of the operational area. Because this tool is intended for use in civilian-centric operations in particular, political boundaries as reference points for working with host-nation

81 Ostrom, Elinor, "A diagnostic approach for going beyond panaceas," *Proceedings of the National Academy of Sciences* 104, no. 39 (2007): 15181-15187.

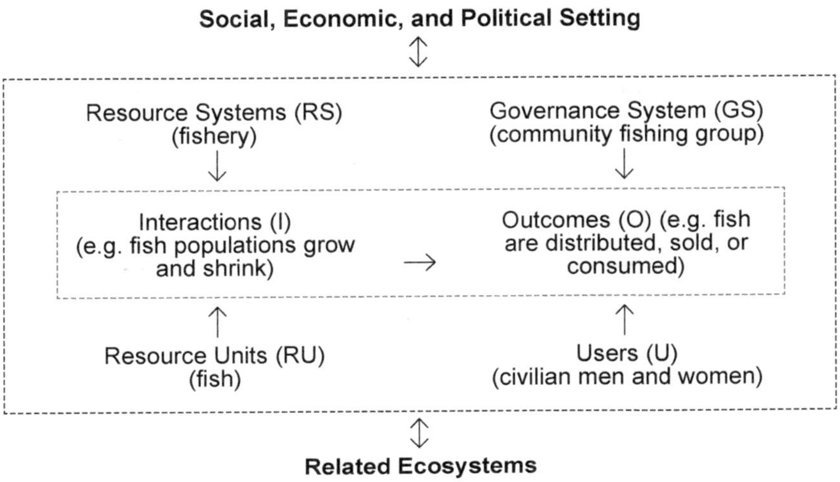

Figure 7.1. Socio-Ecological Systems Approach

governments could provide a useful organising measure. The overlays for a mission area would not be restricted by these boundaries, because they will be illustrated according to the areas for which different data sets were collected, which might not coincide with political lines on the map.

Criteria for each of the three risk factors for the model would be used to populate the respective overlays. Each risk factor would have four or five criteria, assessed by quantifiable gradients. Tables 7.1, 7.2, and 7.3 set out a notional list of criteria that could be used to construct a version of the model for testing. A literature review of gender inequality, armed conflict, and climate change shows that each of these criteria have had data collected against them and have been analysed and have been found relevant to explaining the impacts of these risk factors on human security. Until the model is tested, however, it is not possible to be more definitive as to an individual criterion's fit to the model.

The Gender Inequality Overlay

For purposes of illustration, and because of the significance of gender analysis to the model, it is useful to look in detail at how the notional criteria would be constituted for the Gender Inequality Overlay and the research

1.1 Education Deficit	Average years of schooling (AYS) for women measured at 25 years of age
1.2 Labour Force Participation	Ratio of per cent of male and female labour force participants between ages 14–65
1.3 Maternal Mortality Rate	Number of maternal deaths for every 100,000 live births
1.4 Gender-Differentiated Literacy Rate	Per cent difference of literacy between males and females age 15 and older
1.5 Cultural Violence Against Women	Percentage of women charged with assault who use self-defence plea

Table 7.1. Gender Inequality Criteria

2.1 Physical Displacement	Percent displacement of local people compared to previous year (increase or decrease)
2.2 Fatality Incidence	Number of fatalities in region per year due to armed conflict
2.3 Exposure to Conflict	Months since last exposure to conflict
2.4 Child Malnutrition Status	Difference between regional height-for-weight average and national average
2.5 Access to Nutritionally Dense Food	Annual difference in regional mean cost in reserve currency units ($, £, €) per nutritional unit (calories)

Table 7.2. Armed Conflict Criteria

3.1 Arable Land Cover	Percentage of region covered in vegetation
3.2 Excessive Heat Exposure	Difference between the current monthly average heat index and the historical heat range index
3.3 Climate-Related Disease Incidence	Cumulative deaths over a three-month rolling period for children age five and under induced by high diarrheal incidence
3.4 Storm-Related Damage Costs	Regional expenditures in resrve currency units ($, £, €) spent to manage storm damage

Table 7.3. Climate Change Criteria

that supports the criteria as likely being relevant.[82] At the end of this book, *Appendix 1: Notional Gender Inequality Overlay Criteria Intervals* goes into greater detail in proposing notional gradations within each criterion. This layer focuses on the health and safety of women and girls. The criteria proposed are meant to show general relationships between men/boys and women/girls in a given society (the level of gender inequality) in order to illustrate where and in what ways women may be at risk compared to men. As with any gender analysis, those using this model would need to recognize that there are cultural nuances that need to be considered when determining gender inequality, the likelihood of that inequality affecting an operation, and the appropriate intervention.

Criterion 1.1: Average Educational Deficit

There are many factors that play into the disparity in the educational attainment of boys versus girls, which varies greatly within and between countries. Some of these factors include distance to school, cost of education and supplies, living in rural versus urban areas, and cultural values placed on women and girls. These factors drastically impact the average years of schooling attained by women and girls. Research conducted on the average years of schooling (AYS) in regions of Turkey determined that based on calculations of the Education Gini Index, as the average years of schooling increased, gender-based educational inequality decreased; until approximately six average years of schooling.[83] A similar study done in Tunisia indicates a very similar trend, they found that for every 2.90 additional AYS, gender inequality decreased by 31.1 per cent.[84] Thus, AYS is an apt measurement of inequality.

Criterion 1.2: Labour Force Participation

As reflected in the UN's Gender Inequality Index (GII), participation in the formal labour force traditionally results in net monetary gain, decreasing

---

82 The methodology for establishing the gradients of the criteria for Armed Conflict and Climate Change would be largely the same as that used for the Gender Inequality overlay. As with the data that would populate the overlays, it is important that that it come from reliable studies that have preferably been peer-reviewed, and for which the results evidence statistical rigor.

83 Ekber Tomul, "Measuring regional inequality of education in turkey: An evaluation by Gini index. *Procedia-Social and Behavioral Sciences* 1.1 (2009): 949-952.

84 Salwa Trabelsi, "Regional inequality of education in Tunisia: An evaluation by the Gini Index," *Région et Développement*, no. 37-2013 (2013) 98-117.

the risk of innumerable poverty-related risks for women and girls. There are several reasons that workforce participation decreases gender inequality. First, the freedom to willingly migrate, both to flee from unsafe conditions and to seek employment, is shaped by socioeconomic status.[85] Concurrently, the widespread migration of men to urban areas to diversify household incomes places a disproportionate burden on women, as they become heads of their households and take on greater workloads due to a result of tradition and existing barriers to economic security. In times of conflict, the roles of women as caretakers amplifies women's financial dependence on their male family members,[86] while the unpaid work intensity of women and adolescent girls become significantly higher.[87] Workforce participation, therefore, represents many aspects of gender inequality.

Criterion 1.3: Maternal Mortality Ratio (MMR)

Maternal mortality is an important criterion to determine levels of gender inequality because most maternal deaths are preventable when women have access to adequate health care. A maternal death includes death while pregnant, death within 42 days after the termination of a pregnancy, and / or any death occurring related to pregnancy or a cause that was exacerbated by pregnancy.[88] Maternal mortality is measured by the ratio of maternal deaths related to the number of live births. It is rare that developed countries have an MMR of over 20, and levels of general gender inequality determined by a collection of factors increase as maternal mortality increases.[89] The greatest maternal mortality ratios are in Sub-Saharan Africa and reach well over 1,000 maternal deaths per 100,000 live births. These bracket criteria were used to devise the classes above, showing the clear link between gender equality and MMR.

---

85 Smith *et al.*, "*The Climate-Gender-Conflict Nexus.*"

86 Anita Perešin, Melisa Hasanović, and Kujtim Bytyqi, "Female Returnees from Syria to the Western Balkans," *Perspectives on Terrorism* 15, no. 5 (2021): 29–45.

87 Kati Schindler, "Who Does What in a Household after Genocide? Evidence from Rwanda," DIW Berlin Discussion Paper no. 1072 (November 2010), https://ssrn.com/abstract=1710483.

88 World Health Organization. Indicator Data Registry List: Maternal deaths. The Global Health Observatory. https://www.who.int/data/gho/indicator-metadata-registry/imr-details/4622.

89 UN Development Program, Human Development Reports, "Gender Inequality Index (GII)," (2019), https://www.hdr.undp.org/en/content/gender-inequality-index-gii/

Criterion 1.4: Gender-Differentiated Literacy Rates

Globally, women account for around two-thirds of illiterate adults. In some areas, more than 85 per cent of women are illiterate,[90] and the number of illiterate adults has increased by 37 per cent in sub-Saharan Africa since the 1990s.[91] A strong correlation has been shown between higher illiteracy and intimate partner violence,[92] and each year of school women take has been correlated with up to a 10 per cent reduction in child mortality.[93] A series of gradients would illustrate the per cent difference in illiteracy between males and females aged 15 and older.

Criterion 1.5: Gendered Cultural Violence

Culturally accepted violence against women (for example, domestic abuse, dowry-related homicides) is a difficult metric to measure across cultures due to cultural biases against tracking such information. Therefore, it is possible that a number of the women and adolescent girls charged with assault using the self-defence plea remain or were in positions of surviving domestic or culturally acceptable physical abuse from the plaintiff (or plaintiff equivalent). If such is the case, a metric that takes the per cent of women charged as the defendant (or defendant equivalent) in assault cases pleading self-defence is a potential proxy for the current risk of physical and/or cultural violence against women. This potential metric assumes the region of operation's legal system allows for the recognition of a "self-defence plea" or its cultural equivalent, and thus must be used in the right context among differing regions.

---

90 David W. Hornbeck and Lester M. Salamon, *Human capital and America's future: an economic strategy for the'90s* (Baltimore, MD: Johns Hopkins University Press, 1991).
91 Ibrahim Niankara, "Gender Inequality in Literacy Status and Households Economic Well-Being in Burkina Faso: A Semiparametric Bi-variate Sample Selection Modeling," *International Journal of Economics and Business Research* 17, no. 2 (2018).
92 Rahman, Mosiur, Keiko Nakamura, Kaoruko Seino, and Masashi Kizuki, "Does gender inequity increase the risk of intimate partner violence among women? Evidence from a national Bangladeshi sample,". *PloS ONE* 8, no. 12 (December 2013).
93 Judith Shabaya and Kwadwo Konadu-Agyemang, "Unequal access, unequal participation: some spatial and socio-economic dimensions of the gender gap in education in Africa with special reference to Ghana, Zimbabwe and Kenya," *Compare: A Journal of Comparative Education* 34, no. 4 (September 2004).

Summary

When taken as a composite for an overlay, these individual criteria stack to show the overall current risk to human security that each factor poses. Thus, the units of measurement for each of the criteria need not bear any relationship to one another. The model is a spatial indicator of regionally compounding risks to human security. Each of these criteria will have according shading on the operational map, which when viewed together would show contrasting risks across a region. By delving into detail on one of the three model layers, readers may understand the breadth and flexibility of the proposed approach. The appendix at the back of the book sets out for the proposed gender inequality overlay notional criteria intervals.

Conclusion

This chapter has explored the relationships between armed conflict, gender inequality, and climate change using a novel approach. It sheds light on the linkages between climate change and children as a subpopulation of vulnerable civilians in armed conflicts, and a sketch of the gendered differences that might result. NATO's method has its role in operational analysis, but an SES approach takes different and complementary information into account when performing an analysis for civilian-focused operations, especially when dealing with intersectionalities. From a practical perspective, the stacked overlay approach allows analysts leeway to develop their own new, operationally relevant metrics that can be easily swapped out for the proposed suite of variables to analyse varying conditions in varying locations.

On this note, it is important to recognize that for many, if not most missions, the impacts of climate change on the deploying force itself may not be operationally relevant. Forces train to deploy to operational areas marked by very different climates than the ones in which they are ordinarily based – the prevailing weather conditions are simply an operational fact. What should register in an operational analysis in a civilian-centric operation, however, is the potential compounding impact climate change might be having on at-risk population cohorts such as women and girls.

Two recent studies demonstrate why an in-depth consideration of the complex relationships in the climate change/armed conflict dyad is not necessary for functioning of the model we propose. The first study, by

Mach *et al.* from 2019, is particularly helpful because it assessed structured judgments by a large number of experts from different disciplines as to the relationships between climate change and armed conflict.[94] In general, the experts agreed that during the last 100 years climate change factors had "affected organized armed conflict within countries," and "other conflict drivers are much more influential for conflict risk" than climate change.[95]

The second study, by Ide *et al.* in 2020, looked at instances in which armed conflict began immediately after climate-related disasters. Although the study's authors acknowledged the findings of the Mach study, they assessed that "[t]he literature thus remains divided and little consensual knowledge on disaster-conflict links exists."[96] In looking at a large number of intrastate conflicts, the researchers could find no generic relationship between the climate-related disasters and ensuing armed conflict.[97] They did find that in cases where the disasters occurred in countries with large populations experiencing ethnic exclusion and low levels of human development there was an increased risk of conflict onset in the wake of the disaster.[98] Importantly though, they noted that it did not appear the disaster had caused the violence, instead it appeared that armed groups took advantage of the disasters to initiate hostilities.[99]

Taken together, the results of these studies indicate that commanders and planners should not ignore the possibility that climate change could have operational impacts on a mission area's population. The results also suggest that in most conflicts any causal influence climate has is most likely indirect and difficult to meaningfully measure separately, which is why the authors developed an integrated model with flexible components. This model could be a practical tool for commanders, planners, and operators, but it would need to be tested and iterated upon to meaningfully vet its usefulness. Ideally, if used consistently, the tool would save staff time and incorporate significant data not normally analysed during civilian-focused operations. Importantly, this chapter has emphasized that women and children in the context of armed conflict and a more extreme climate deserve special attention in operations. This special attention can not only

94 Mach *et al.*, "Climate as a risk factor," 193.
95 Ibid., 194.
96 Tobias Ide, Michael Brzoska, Jonathan F. Donges, and Carl-Friedrich Schleussner, "Multi-method evidence for when and how climate-related disasters contribute to armed conflict risk," *Global Environmental Change* 62 (2020): 2.
97 Ibid., 7.
98 Ibid., 7.
99 Ibid., 7.

save civilian and military personnel time, resources, and potentially their lives, it can help make WPS in military operations a reality.

References

Akresh, Richard, Leonardo Lucchetti, and Harsha Thirumurthy. "Wars and child health: Evidence from the Eritrean-Ethiopian Conflict." *Journal of Development Economics* 99, no. 2 (2012): 330-340.

Akresh, Richard, Philip Verwimp and Tom Bundervoet, "Civil War, Crop Failure, and Child Stunting in Rwanda." *Economic Development and Cultural Change* 59, no. 4 (2011): 777-810.

Alam, Mayesha. "A cross-cutting agenda: Gender, climate change and conflict." *ECDPM Great Insights Magazine* 8, no. 4 (Autumn 2019), https://ecdpm.org/great-insights/complex-link-climate-change-conflict/cross-cutting-agenda-gender-climate-change-conflict/.

Alderman, Harold, Jere R. Behrman, Victor Lavy, and Rekha Menon. "Child health and school enrollment: A longitudinal analysis." *The Journal of Human Resources*, 36(1) (2001): 185–203.

Berg, Annika. "Gender analysis—Planning Phase 2," Nordic Centre for Gender in Military Operations, Powerpoint Presentation (undated, but prior to March 2020) (copy on file with author).

Brown, Molly E., David Backer, Trey Billing, Peter White, Kathryn Grace, Shannon Doocy, and Paul Huth. "Empirical studies of factors associated with child nutrition: highlighting the evidence about climate and conflict shocks." *Food Security* 12(5) (2020): 1241-1252.

Brown, Molly E., Kathryn Grace, Trey Billing, and David Backer. "Considering climate and conflict conditions together to improve interventions that prevent child acute malnutrition." *The Lancet Planetary Health*, 5(9) (2021): 654–658.

Brzoska, Michael. "Climate Change As A Risk Amplifier: On the Links Between Climate Change and Conflict." *Ethics and Armed Forces, Controversies in Peace Ethics and Security Policy* 1 (2021): 10–17.

Buhaug, Halvard and Nina von Uexkull, "Vicious Circles: Violence, Vulnerability, and Climate Change." *Annual Review of Environment and Resources* 46 (2021): 545-68.

Bundervoet, Tom, Phillip Verwimp, and Richard Akresh. "Health and Civil War in rural Burundi." *Journal of Human Resources*, 44(2) (2009): 536–563.

Butler, Judith. "Why is the idea of 'gender' provoking backlash the world over?" *The Guardian*, 23 October 2021, https://www.theguardian.com/us-news/commentisfree/2021/oct/23/judith-butler-gender-ideology-backlash.

Buvinić, Mayra, Monica Das Gupta and Olga N. Shemyakina. "Armed Conflict, Gender and Schooling." *The World Bank Economic Review* 28, no. 2 (2014): 311-319. https://www.theguardian.com/us-news/commentisfree/2021/oct/23/judith-butler-gender-ideology-backlash

Bzeouicha, Boudour. "Microfinance and The Improvement of Social Welfare of Beneficiary Households: Evidence from Tunisia." *Biltürk Journal of Economics and Related Studies* 1(3), (2019): 188-205.

Caldwell, Jeff. "Hot Nights Rob Corn Yields." *Successful Farming* (21 June 2012), https://www.agriculture.com/news/crops/hot-nights-rob-cn-yields_2-ar24826.

Chhay, Daraka. (2011). "Women's economic empowerment through microfinance in Cambodia." *Development in Practice* 21(8) (2011): 1122-1137.

Cohen, Marc J. and Per Pinstrup-Andersen. "Food Security and Conflict." *Social Research*, 66(1) (1999): 375–416.

Collier, Paul and Anke Hoeffler. "Post-Conflict Risks." *Journal of Peace Research* 45(4) (2008): 461-478.

Craig, Marlies H., Robert W. Snow, and David le Sueur. "A climate-based distribution model of malaria transmission in Sub-Saharan Africa." *Parasitology Today* 15(3) (1999): 105–111.

Cumming, G.S., G. Barnes, Stephen M. Perz, M. Schmink, Kathryn E. Sieving, J. Southworth, Michael W. Binford, R.D. Holt, C. Stickler, and T.V. Holt. "An Exploratory Framework for the Empirical Measurement of Resilience." *Ecosystems* 8 (November 2005): 975-987.

Cushing, Laura, Rachel Morello-Frosch, and Alan Hubbard. "Extreme heat and its association with social disparities in the risk of spontaneous preterm birth." *Paediatric and Perinatal Epidemiology* 36, no. 1 (January 2022): 13–22.

Dankelman, Irene (ed.). Gender and climate change: An introduction. London: Routledge, 2010.

Dharmapuri, Sahana. "Just Add Women and Stir?" *Parameters* 41, no. 1 (2011): 56–70.

Dionicio López, Carlos Francisco, Neora Alterman, Ronit Calderdon-Margalit, Michael Hauzer, Itai Kloog, and Ranaan Raz, "Postnatal exposure to ambient temperature and rapid weight gain among infants delivered at term gestations: a population-based cohort study." *Paediatric and Perinatal Epidemiology* 36, no. 1 (January 2022): 26–35.

Duque, Valentina. "Early-life conditions and child development: Evidence from a violent conflict." *SSM—Population Health* 3 (2017): 121-131.

Enzinna, Wes. "A Dream of Secular Utopia in ISIS' Backyard." *New York Times Magazine*, 24 November 2015, https://www.nytimes.com/2015/11/29/magazine/a-dream-of-utopia-in-hell.html.

Frounfelker, Rochelle L., Nargis Islam, Joseph Falcone, Jordan Farrar, Chekufa Ra, Cara M. Antonaccio, Ngozi Enelmah and Therese S. Betancourt. "Living through war: Mental health of children and youth in conflict-affected areas." *International Review of the Red Cross* 101, no. 911 (2019): 481-506.

Fukuda-Parr, Sakiko. "What does feminization of poverty mean? It isn't just lack of income." *Feminist Economics* 5, no. 2 (1999): 99-103.

Gallopín, Gilberto C. "Linkages between vulnerability, resilience, and adaptive capacity." *Global Environmental Change* 16, no. 3 (2006): 293-303.

Gardam, Judith and Michelle Jarvis. *Women, Armed Conflict and International Law*. The Hague: Kluwer Law International, 2001.

Glewwe, P., Jacoby, H. G., & King, E. M. Early childhood nutrition and academic achievement: A longitudinal analysis. *Journal of Public Economics* 81(3), 345–368, 2001. https://doi.org/10.1016/s0047-2727(00)00118-3.

Goh, Amelia H. X. "A Literature Review of the Gender-Differentiated Impacts of Climate Change on Women's and Men's Assets and Well-Being in Developing Countries" CAPRi Working Paper no. 106 (2012), http://cdm15738.contentdm.oclc.org/utils/getfile/collection/p15738coll2/id/127247/filename/127458.pdf.

Guanti, Fredah and Yoko Kijima. "Ethnic Violence and Birth Outcomes: Evidence from Exposure to the 1992 Conflict in Kenya." *Demography* 57 (2020): 423-444.

Headquarters, Department of the Army (HQDA). *Field Manual 3-57, Civil Affairs Operations.* Washington, D.C.: HQDA, 2021, https://armypubs.army.mil/ProductMaps/PubForm/Details.aspx?PUB_ID=1022687.

Heft-Neal, Sam, Anne Driscoll, Wei Yang, Gary Shaw, and Marshall Burke. "Associations between wildfire smoke exposure during pregnancy and risk of preterm birth in California." *Environmental Research* 203 (2022): 111872, 1-10.

Hillenbrand, Emily. "Transforming Gender in Homestead Food Production." *Gender & Development* 18, no. 3 (2010): 411–25.

Holling, C. S. "Resilience and stability of ecological systems." *Annual Review of Ecology and Systematics* 4 (November 1973): 1-23.

Holling, C. S. and Gary K. Meffe. "Command and Control and the Pathology of Natural Resource Management." *Conservation Biology* 10, no. 2 (1996): 328-337.

Hornbeck, David W. and Lester M. Salamon (eds.). *Human capital and America's future: an economic strategy for the'90s.* Baltimore, MD: Johns Hopkins University Press, 1991.

Ide, Tobias, Michael Brzoska, Jonathan F. Donges, and Carl-Friedrich Schleussner. "Multi-method evidence for when and how climate-related disasters contribute to armed conflict risk." *Global Environmental Change* 62 (2020): 1–8.

International Committee of the Red Cross. "Seven things you need to know about climate change and conflict," 9 July 2020, https://www.icrc.org/en/document/climate-change-and-conflict.

Jegosothy, Edward, Deborah A. Randall, Jane B. Ford, Tanya A. Nipita, and Geoffrey G. Morgan. "Maternal factors and risk of spontaneous preterm birth due to high ambient temperatures in New South Wales, Australia." *Paediatric and Perinatal Epidemiology* 36, No. 1 (January 22): 4-12.

Jones, Ann. War is Not Over when It is Over—Women Speak Out from the Ruins of War. New York: Macmillan Publishers, 2010.

Kamarck, Elaine. "The challenging politics of climate change," Brookings Institute (23 September 2019), https://www.brookings.edu/research/the-challenging-politics-of-climate-change/.

Kumar, Santosh, Ramona Molitor, and Sebastian Vollmer. "Drought and Early Child Health in Rural India." *Population and Development Review* 42, no. 1 (2016): 53-68.

Mach, Katherine J., Caroline M. Kraan, W. Neil Adger, Halvard Buhaug, Marshall Burke, James D. Fearon, Christopher B. Field, *et al.*, "Climate as a risk factor for armed conflict." *Nature* 571 (2019): 193-97.

March, Candida, Ines Smyth, and Maitrayee Mukhopadhyay. *A Guide to Gender-Analysis Frameworks*. Oxford: Oxfam, 1999.

McLanahan, Sara S. and Erin L. Kelly. "The feminization of poverty." In *Handbook of the Sociology of Gender*, edited by Janet Saltzman Chafetz, 127-145. Boston, MA: Springer, 2006.

Miller, Korin. "How Pregnant Women and Babies Can Stay Safe in Wildfire Smoke," *Whattoexpect.com* (17 September 2020), https://www.whattoexpect.com/news/family/wildfire-smoke-safety-pregnant-women-babies.

Moser, Caroline. Gender Planning and Development: Theory, Practice and Training (1st ed.). London: Routledge, 1993.

Myrrtinen, Henri. "Drowning in complexity? Preliminary Findings on Gender, Peacebuilding and Climate Change in Honduras." In *Addressing Global Environmental Challenges from a Peace Ecology Perspective*. edited by Hans Günter Brauch *et al.*, The Anthropocene: Politik—Economics—Society—Science 4, 97-111. Cham, Switzerland: Springer International Publishing, 2016.

NATO, Allied Command Operations. *Gender Functional Planning Guide*. Brussels: Supreme Allied Headquarters Europe, July 2015.

NATO, Allied Command Operations and Allied Command Transformation. *Bi-SC Directive 040-001, Integrating UNSCR 1325 and Gender Perspective into the NATO Command Structure*. Brussels and Norfolk, VA: Supreme Allied Headquarters Europe and Supreme Allied Command Transformation, 17 October 2017.

NATO, Allied Command Operations (ACO). *Comprehensive Operations Planning Directive Interim Version 2.0*. Mons, Belgium: ACO, 4 October 2013.

Niankara, Ibrahim. "Gender Inequality in Literacy Status and Households' Economic Well-Being in Burkina Faso: A Semiparametric Bi-variate Sample Selection Modeling." International Journal of Economics and Business Research 17, no. 2 (2018): 218-242.

O'Rourke, Ronald. "Renewed Great Power Competition: Implications for Defense—Issues for Congress." *Congressional Research Service* (21 December 2021): 1–15, R43838.pdf (fas.org).

Ostrom, Elinor. "A diagnostic approach for going beyond panaceas." *Proceedings of the National Academy of Sciences* 104, no. 39 (2007): 15181-15187.

Park, Bo Young, Ian Boles, Samira Monavvari, Shivani Patel, Arriel Alvarez, Mie Phan, Maria Perez, and Ruofan Yao. "The association between wildfire exposure in pregnancy and foetal gastroschisis: A population-based cohort study." *Paediatric and Perinatal Epidemiology* 36, no. 1 (January 2022): 45-53.

Patil, Anil. "Why caregiving needs to be 'de-feminized,'" White Swan Foundation (28 March 2018), https://www.whiteswanfoundation.org/caregiving/how-the-role-of-caring-disproportionately-burdens-women.

Perešin, Anita, Melisa Hasanović, and Kujtim Bytyqi. "Female Returnees from Syria to the Western Balkans." *Perspectives on Terrorism* 15, no. 5 (2021): 29–45.

Prescott, Jody M., Robin Lovell, and Team Sandhurst (Cameron Chambers, Emma Miner, Cyrus Oswald, Emily Smith). "Gender inequality, armed conflict and climate change: why militaries can and should map compounded risk." International Committee of the Red Cross, *Humanitarian Law & Policy Blog* (29 September2022), https://blogs.icrc.org/law-and-policy/2022/09/29/gender-inequality-armed-conflict-climate-change-militaries-map-compounded-risk/.

Prescott, Jody M. "Gender Blindness in US Doctrine." *Parameters* 50, no. 4 (2020): 21-32.

Prescott, Jody M. *Armed Conflict, Women and Climate Change*. London: Routledge, 2019.

Prescott, Jody M. "Lawfare: Softwiring the Network." *The Three Swords* 32 (2017): 13-18.

Price, Roz. "Climate Change as a driver of conflict in Afghanistan and other Fragile and Conflict Affected States." *KD4 Helpdesk Report*, Institute of Development Studies (18 January 2019), 1–18, https://assets.publishing.service.gov.uk/media/5ca20ba940f0b625df8d85f1/527_Climate_change_as_a_driver_of_conflict_in_Afghanistan_and_other_FCAS.pdf.

Rahman, Mosiur, Keiko Nakamura, Kaoruko Seino, and Masashi Kizuki. "Does gender inequity increase the risk of intimate partner violence among women? Evidence from a national Bangladeshi sample." *PloS ONE* 8, no. 12 (2013): 1-12.

Requia, Weeberb J. Heresh Amini, Matthew D. Adams, and Joel D. Schwarz. "Birth weight following pregnancy wildfire smoke exposure in more than 1.5 million newborns in Brazil: A nationwide case-control study." *The Lancet Regional Health, Americas* (14 March 2022), https://www.thelancet.com/journals/lanam/article/PIIS2667-193X(22)00046-1/fulltext .

Sheffield, Perry E. and Phillip J. Landrigan. "Global Climate Change and Children's Health: Threats and Strategies for Prevention." *Environmental Health Perspectives* 119, no. 3 (2011): 291-298.

Sivin, Joshua Graff and Jeffrey Shrader. "Temperature Extremes, Health, and Human Capital." *The Future of Children* 26, no. 1 (2016): 31-50.

Smith, Jessica M., Lauren Olosky, and Jennifer Grosman Fernández, "The Climate-Gender-Conflict Nexus – Amplifying women's contributions at the grassroots," Georgetown Institute for Women, Peace and Security, (2021): 5-14 (report), https://giwps.georgetown.edu/wp-content/uploads/2021/01/The-Climate-Gender-Conflict-Nexus.pdf .

Speranza, Chinwe Ifejika and Edward Bikketi. "Engaging with Gender in Water Governance and Practice in Kenya." In *Water Security Across the Gender Divide*, edited by C. Fröhlich, G. Gioli, R. Cremades, and H. Myrttinen, 125-150. Cham, Switzerland: Springer International Publishing, 2018.

Tax, Meredith. *A Road Unforeseen — Women Fight the Islamic State*. New York: Bellevue Literary Press, 2016.

Terry, Geraldine. *Climate change and gender justice*. Rugby, UK: Practical Action Publishing, 2009.

Tomul, E. Measuring regional inequality of education in turkey: An evaluation by gini index. *Procedia-Social and Behavioral Sciences* 1.1 (2009): 949-952.

Trabelsi, Salwa. "Regional inequality of education in Tunisia: An evaluation by the Gini Index." *Région et Développement*, no. 37 (2013): 98-117, https://regionetdeveloppement.univ-tln.fr/wp-content/uploads/5_Trabelsi-1.pdf .

Tranchant, Jean-Pierre, Patricia Justino, and Cathérine Müller, "Political Violence, Drought and Child Malnutrition: Empirical Evidence from Andhra Pradesh, India." HiCN Working Paper no. 173, Households in Conflict Network, The Institute of Development Studies (University of Sussex) (April 2014): 1–52, https://www.hicn.org/wp-content/uploads/sites/10/2012/06/HiCN-WP-173.pdf .

Tsujimoto, Takahiro and Yoko Kijima. "Effects of conflict on child health: Evidence from the 1990-19994 Northern Mali Conflict." *Health Economics* 29 (2020): 1456-1474.

UN Security Council Resolution 1325, On Women, Peace and Security, UN DOC. S/RES/1325 (31 October 2000).

UN Standing Committee on Nutrition. Sixth report on the world nutrition situation: Progress in Nutrition (2010).

Valente, Christine. "Education and Civil Conflict in Nepal." *The World Bank Economic Review* 28, no. 2 (2014): 354–383.

World Health Organization (WHO). Indicator Data Registry List: Maternal deaths. The Global Health Observatory. https://www.who.int/data/gho/indicator-metadata-registry/imr-details/4622.

8

BRINGING WOMEN, PEACE AND SECURITY ONLINE

Applications in the Cyber Domain

Alexis Henshaw

This chapter explores the application of the Women, Peace and Security (WPS) agenda in the cyber domain. In doing so, it examines how technology and online security issues interface with the pillars of WPS: Participation, Protection, Prevention, Relief and Recovery, and Institutionalization, which is sometimes regarded as a fifth pillar of the agenda.[1] The key argument of this chapter is that cyberspace has largely been overlooked as a source of gendered security threats. The rise of gender-based extremism (especially far-right, misogynist, and/or conspiracy-based beliefs) is one significant challenge that directly corresponds to the mandates of UN Security Council resolutions on WPS. The COVID-19 pandemic, which has forced many individuals to conduct more of their communications online, is another challenge with significant and gendered impacts. At the same time, looking at possible applications of the WPS agenda in cyberspace can also prompt a re-examination of "cybersecurity" and related concepts. Specifically, the framework of WPS calls for the placement of women and communities (not only states and state systems) at the center of policymaking.

1 Mary K. Meyer McAleese, "WPS and the Organization of American States," in *The Oxford Handbook of Women, Peace, and Security*, ed. Sara E. Davies and Jacqui True (Oxford; New York: Oxford University Press, 2019), 413–27.

The Cyber Domain, Revisited

Approaching the study of cybersecurity with a gender lens entails developing a working definition of "cybersecurity" and the "cyber domain." Though cybersecurity is often viewed narrowly as the protection of computing systems and infrastructure, the functional definition of the "cyber domain" as used by security forces (including NATO and the U.S. Department of Defense) is more expansive. These agencies see the cyber domain as encompassing computer networks and infrastructure, plus information and communication networks more broadly (for example, telecommunications networks).[2] The remit of cybersecurity entails not only the protection of these networks but also the tasks of ensuring access and upholding established international law and norms in these spaces. Experts broadly describe at least three (and possibly four) main threats as the referent of cybersecurity. These are cybercrime, cyberconflict or cyberwar, and cyber espionage.[3] In addition, commentators have noted that cybersecurity also overlaps with the space of information operations.[4] This creates an imperative for cyber operations to address disinformation campaigns, propaganda, and the use of technology by extremist groups when these campaigns are conducted using information and communications technology.

The notion of a cyber "domain" is not uncontested. Commentators have pointed out the ways in which cyber fundamentally differs from other, physical domains of warfighting. They argue that the designation of cyberspace as a domain is artificially narrow, representing an effort to bound or perhaps establish sovereignty over a space that is purely metaphorical.[5] Because much of the architecture that supports cyberspace is privately owned, and because experts have played an outsized role in

2 Glenn Alexander Crowther, "The Cyber Domain," *The Cyber Defense Review*, Fall 2017 (2017): 63–78; NATO, "Cyber Defence Pledge," NATO, July 8, 2016, https://www.nato.int/cps/en/natohq/official_texts_133177.htm; U.S. Cyber Command, "Mission and Vision," U.S. Cyber Command, n.d., https://www.cybercom.mil/About/Mission-and-Vision/.

3 Mark Raymond, "Managing Decentralized Cyber Governance," in *Understanding Cyber Security: Emerging Governance and Strategy* (New York, N.Y: Rowman & Littlefield, 2018), 25–48.

4 Crowther, "The Cyber Domain."

5 Michael V. Hayden, "The Future of Things Cyber," in *Understanding Cyber Security: Emerging Governance and Strategy* (New York, N.Y: Rowman & Littlefield, 2018), xi–xv; Michael Kreuzer, "Cyberspace Is an Analogy, Not a Domain: Rethinking Domains and Layers of Warfare for the Information Age," The Strategy Bridge, July 8, 2021, https://thestrategybridge.org/the-bridge/2021/7/8/cyberspace-is-an-analogy-not-a-domain-rethinking-domains-and-layers-of-warfare-for-the-information-age.

Country	No. of Mentions of Cybersecurity
Ireland (2019)	3
Italy (2014)	1
Kenya (2020)	1
Namibia (2019)	5
The Netherlands (2021)	1
The Philippines (2009)	1
South Korea (2014)	1

Table 8.1. Countries with NAPs Referencing Cybersecurity
Source: Hamilton and Shepherd (2020).

developing governance structures, the attempts to assert state supremacy or sovereignty in this space are far from clear-cut. Critiques like these offer an entry point for questions about the securitization and militarization of cyberspace. What if we were to place communities and users at the center of cybersecurity, or at least make such entities co-equal with systems and state interests?

Given the paradigm shift this entails, we begin to see where WPS can enter the conversation. First, WPS is a comprehensive and binding legal framework at the international level for holding state actors accountable for inclusive security. Encompassing, to date, 10 United Nations Security Council Resolutions, the WPS Agenda has also spawned several dozen National Action Plans (NAPs) and implementation plans from international organizations including NATO. Though the WPS framework has been far from a perfect tool, it is a living agenda that continues to evolve. It therefore provides a useful entry point for engaging discussions on new, gendered security threats.

Related to this, there is evidence that digital and cybersecurity issues have already come to WPS. In their analysis of National Action Plans, Hamilton and Shepherd find seven countries that have incorporated some mention of cyber- in their NAPs. Table 8.1 highlights these countries. Some of these mentions do not equate to substantive discussion.[6] Some states, like Italy and The Philippines, mention cybersecurity in early NAPs, but omit reference to the issue on subsequent revisions.

6 For example, the NAP from The Philippines mentions cyber issues once, compared to 14 mentions each for sexual violence and refugees. Caitlin Hamilton and Laura J. Shepherd, "WPS National Action Plans: Content Analysis and Data Visualization, V2," 2020, http://www.wpsnaps.org.

Still, some states are seeking to pioneer norms of inclusive cybersecurity through their National Action Plans or through national cybersecurity strategies. Ireland's most recent NAP calls for efforts to engage more women as cybersecurity professionals and to mainstream gender perspectives in the practice of cybersecurity from 2020 onward.[7] Namibia's NAP calls for developing responses to gender-based cybercrime and devoting more attention to the gendered impact of cybercrime and cybersecurity issues.[8] The UK is an example of a country that has referenced gender in its National Cyber Strategy. The 2022 strategy promotes gender mainstreaming in three specific areas: 1) The need to diversify the labour pool in the technology sector; 2) the need to highlight gender-based online violence; and 3) the need to promote inclusive design and use of online spaces.[9] Taken together, these plans offer insight into emerging norms in this area.

A final justification for engaging in a discussion about digital insecurity through the framework of WPS is that technology has become so deeply intertwined with many other referent issues for the agenda. This chapter demonstrates how information and communications technology (ICT) platforms and other forms of technology are engaged in the perpetration of sexual violence, radicalization, political violence, the targeting of LGBTQI+ populations, and response to crisis and humanitarian emergencies. If the WPS agenda is to continue to address inclusion as outlined in resolutions and National Action Plans, it must engage in dialogue about the uses of technology. The discussions below offer a brief overview of possible applications and the relationship of WPS to military operations in the four pillars of participation, protection, prevention, and relief and recovery. This chapter closes with some comments and reflections relevant to the institutionalization of WPS in the cyber domain.

7 Government of Ireland, "Women, Peace and Security: Ireland's Third National Action Plan for the Implementation of UNSCR 1325 and Related Resolutions" (Dublin: Government of Ireland, 2019), https://www.dfa.ie/media/dfa/ourrolepolicies/womenpeaceandsecurity/Third-National-Action-Plan.pdf.

8 Republic of Namibia, "Namibia National Action Plan on Women Peace and Security" (Windhoek, Namibia: Republic of Namibia, 2019), https://www.peacewomen.org/sites/default/files/Namibia%20NAP%20(2019-2024).pdf.

9 HM Government, "National Cyber Strategy 2022" (London, UK: HM Government, 2022), https://assets.publishing.service.gov.uk/government/uploads/system/uploads/attachment_data/file/1053023/national-cyber-strategy-amend.pdf.

Participation

It is no secret that women are underrepresented in the cybersecurity sector, but the impact of this gender imbalance deserves further attention. One industry association concluded in 2019 that the cybersecurity workforce was just 24 per cent women, yet even that figure includes practitioners who are dual-hatted and/or in "non-traditional" cybersecurity roles.[10] Paradoxically, though women in cybersecurity are more likely to pursue professional development they also make less money, are less likely to hold executive positions, and are more likely to report harassment and discrimination.[11] The attrition of women seems to follow the kind of "leaky pipeline" model envisioned in other professions, with women lost at every career stage. While women who study computer science or who are at early career phases often struggle to navigate sexism, stereotyping, and imposter syndrome; more established women can be driven out of technology professions by harassment, concerns over work-life balance, or the lack of role models and mentors.[12]

As cybersecurity becomes a more salient concern, especially for organizations involved in national security, we are only beginning to understand the implications of this exclusion. Training and retaining more women could help address a critical shortage of cybersecurity professionals. Beyond that, scholars and practitioners have argued that a more inclusive cybersecurity workforce could lead to better user experience, more attention

---

10 (ISC)2, "Women in Cybersecurity: Young, Educated, and Ready to Take Charge" (International Information System Security Certification Consortium, Inc., 2019), https://www.isc2.org/-/media/ISC2/Research/ISC2-Women-in-Cybersecurity-Report.ashx?la=en&hash=4C3B33AABFBEAFDDA211856CB274EBDDF9DBEB38. Frost & Sullivan (2017), using a narrower methodology, found that only 11% of cybersecurity professionals were women.

11 Frost & Sullivan, "The 2017 Global Information Security Workforce Study: Women in Cybersecurity" (Clearwater, FL: Center for Cyber Safety and Education/Frost & Sullivan, 2017), https://www.isc2.org/-/media/Files/Research/ISC2-Women-in-Cybersecurity-2017.ashx?la=en&hash=FA78E4BDA2F858A0D3CFAC75AF9E789369A0BC8D; (ISC)2, "Women in Cybersecurity: Young, Educated, and Ready to Take Charge."

12 J. McGrath Cohoon, Monica Nable, and Paul Boucher, "Conflicted Identities and Sexism in Computing Graduate Programs," in *2011 Frontiers in Education Conference (FIE)*, 2011, S1H-1-S1H-5, https://doi.org/10.1109/FIE.2011.6142915; Katrina Falkner *et al.*, "Gender Gap in Academia: Perceptions of Female Computer Science Academics," *Proceedings of the 2015 ACM Conference on Innovation and Technology in Computer Science Education*, 2015, https://dl.acm.org/doi/pdf/10.1145/2729094.2742595?casa_token=OOQi6kGVSwQAAAAA:U7SDNWSTd6c4mcJMmz6JI6sfuAhb2ynpvXhfTLa-cc7jjFyU1-3TbAdIFiD2QzQQN2ZHQyiH42Sa; Reshma Saujani, "Disrupting Tech's Growing Gender Gap" (Qualtrics/Girls Who Code Forum, Online, January 27, 2021).

to data security and privacy issues, and more effective information-sharing about security threats.[13] As the remainder of this chapter shows, women also face distinct forms of cyberthreat and cybercrime. Enhancing women's participation in the cybersecurity sector can call attention to the ways in which online insecurities and crime are gendered.

The participation pillar can further be used to shed light on how technology fosters the inclusion of women in political debate. United Nations Security Council Resolution (UNSCR) 1325 envisioned a system in which women and girls could be full and equal participants in peace processes and post-conflict reconstruction. This is an aim that has yet to be fully realized. Women remain underrepresented as mediators and signatories to peace agreements.[14] At the same time, there are indications that the implementation of gender provisions in peace agreements, for example in Colombia and South Sudan, has failed to meet expectations.[15] More recently, some have questioned why women remain sidelined or confined to "advisory" roles in high-profile peace talks including those stemming from conflicts in Syria, Yemen, and Ukraine.[16] Promoting civil society participation in WPS-related initiatives is important not only to diplomatic peace efforts, but also to the functioning of peacekeeping and post-conflict stability operations.

In previous research, I wrote about how communications technologies, especially social media, can play an important role in supporting peace processes. Looking particularly at the process that culminated in the 2016 peace accords in Colombia, I showed how women representing the Revolutionary Armed Forces of Colombia (FARC) at the talks used

---

13 Nir Kshetri and Maya Chhetri, "Gender Asymmetry in Cybersecurity: Socioeconomic Causes and Consequences," *Computer*, no. February 2022 (2022): 72–77; Camélia Radu and Nadia Smaili, "Board Gender Diversity and Corporate Response to Cyber Risk: Evidence from Cybersecurity Related Disclosure," *Journal of Business Ethics* 177, no. 2 (May 1, 2022): 351–74, https://doi.org/10.1007/s10551-020-04717-9.

14 UN Women, "Women's Meaningful Participation in Negotiating Peace and the Implementation of Peace Agreements" (New York: UN Women, 2018), https://www.unwomen.org/-/media/headquarters/attachments/sections/library/publications/2018/egm-womens-meaningful-participation-in-negotiating-peace-en.pdf?la=en&vs=3047.

15 Heidi Gasperetti *et al.*, "The Pandemic Will Not Stop Us: The Impact of COVID-19 on Women's Peace Activism in Colombia, the Philippines, South Sudan and Ukraine," Case Study Reports (New York, NY: Global Network of Women Peacebuilders/Center for Global Affairs at NYU, March 2021), https://gnwp.org/wp-content/uploads/NYU-Research-Policy-Brief_March-2021.pdf.

16 Cynthia Enloe, *The Big Push: Exposing and Challenging the Persistence of Patriarchy* (Myriad Editions, 2017); Chitra Nagarajan and Azadeh Moaveni, "Another Deeply Gendered War Is Being Waged in Ukraine," March 15, 2022, https://www.aljazeera.com/opinions/2022/3/15/another-deeply-gendered-war-is-being-waged-in-ukraine.

social media to foster public dialogue, to highlight gender issues, and to promote transparency. An important finding of this research was that women were earlier adopters and more successful users (in terms of reach and engagement) than their male counterparts. Women also promoted a distinct variety of issue linkages, including more discussion on women's rights and LGBTQI+ rights issues.[17] We can also see successful deployments of technology to support women's involvement in peace talks in Yemen, where the United Nations recently used online dialogues for information sharing and to collect public opinion. Organizers report that the anonymity provided by the online platform facilitated the involvement of women, while the use of artificial intelligence provided real-time translation among participants using different languages or dialects.[18] Such projects suggest that technology may play an important role in creating inclusive peace.

Research by the Global Network of Women Peacebuilders further highlights the importance of Internet-based and mobile communications networks for women's groups and civil society activists. Especially during the COVID-19 pandemic, these technologies have allowed women to maintain dialogues even where in-person gatherings became impossible.[19] Yet the movement online has come with new challenges. In many parts of the world, women are less likely to have reliable access to online/mobile technologies and are less likely to be experienced in using these platforms.[20] As a result, some have complained that their lack of technological expertise

---

17 Alexis Leanna Henshaw, "'Peace with a Woman's Face': Women, Social Media, and the Colombian Peace Process," Contexto Internacional 42, no. 3 (2020): 515–38.

18 OSESGY, "Cutting-Edge Tech in the Service of Inclusive Peace in Yemen," OSESGY, August 3, 2020, https://osesgy.unmissions.org/cutting-edge-tech-service-inclusive-peace-yemen; OSESGY, "UN Special Envoy's Office for Yemen Concludes First Online Mass Consultations," OSESGY, June 10, 2020, https://osesgy.unmissions.org/un-special-envoy%E2%80%99s-office-yemen-concludes-first-online-mass-consultations.

19 Anne-Marie Buzatu *et al.*, "Women, Peace, and Security and Human Rights in the Digital Age: Opportunities and Risks to Advance Women's Meaningful Participation and Protect Their Rights," Policy Brief (Global Network of Women Peacebuilders/ICT4Peace Foundation, 2021), https://gnwp.org/wp-content/uploads/PolicyBriefGNWP-2021c.pdf; Gasperetti *et al.*, "The Pandemic Will Not Stop Us: The Impact of COVID-19 on Women's Peace Activism in Colombia, the Philippines, South Sudan and Ukraine."

20 See, e.g., research on the digital gender gap from IISD's SDG Knowledge Hub, "What Is the Gender Digital Divide, and Why Should It Matter for the SDGs?," 2017, https://sdg.iisd.org:443/commentary/guest-articles/what-is-the-gender-digital-divide-and-why-should-it-matter-for-the-sdgs/; Human Rights Centre, "Ways to Bridge the Gender Digital Divide from a Human Rights Perspective" (Essex, UK: Human Rights, Big Data and Technology Project at the Human Rights Centre of the University of Essex, n.d.), https://www.ohchr.org/Documents/Issues/Women/WRGS/GenderDigital/HRBDT_submission.pdf; OECD, "Bridging the Digital Gender Divide: Include, Upskill, Innovate" (Paris, France: OECD, 2018), http://www.oecd.org/digital/bridging-the-digital-gender-divide.pdf.

was used as a justification to exclude them from participation.[21] Even where women are able to get online, limited bandwidth and / or unreliable electric grids have sometimes forced them to turn to platforms that are less secure, but easier to access.[22] Some women report facing increased online harassment as a result of their activities online and / or on social media.[23]

There are potential solutions that could promote gendered inclusion in peace and stability operations via ICT. As a baseline, ensuring reliable access to electricity and mobile networks is a necessity, especially in rural areas. Basic training on how to use technologies securely can also be important, as women engaged in peace / post-conflict processes have discussed learning online security best practices only through trial and error during the COVID-19 pandemic.[24] This sort of training may be especially important for older women. While some have reported that the movement of dialogues online has yielded more youth engagement, older women may have lower literacy rates and a lower comfort level with technology.[25] Private actors, including technology companies and mobile service providers, may also have a role to play. Offering free or reduced-cost access to data / Internet plans, for example, could facilitate greater participation among poor and rural women. Women also report needing private spaces in which to carry out online conversations, as many homes may share mobile devices among family members.[26] Offering women advocates low-cost access to additional devices, or private spaces for discussion might address some of these issues. Efforts to improve technology access also illustrate the ongoing importance of literacy efforts, as basic literacy is often necessary to navigate technology.

---

21 Gasperetti *et al.*, "The Pandemic Will Not Stop Us: The Impact of COVID-19 on Women's Peace Activism in Colombia, the Philippines, South Sudan and Ukraine."
22 Buzatu *et al.*, "Women, Peace, and Security and Human Rights in the Digital Age: Opportunities and Risks to Advance Women's Meaningful Participation and Protect Their Rights."
23 Gasperetti *et al.*, "The Pandemic Will Not Stop Us: The Impact of COVID-19 on Women's Peace Activism in Colombia, the Philippines, South Sudan and Ukraine." See also the discussion of protection in the following section.
24 Buzatu *et al.*, "Women, Peace, and Security and Human Rights in the Digital Age: Opportunities and Risks to Advance Women's Meaningful Participation and Protect Their Rights"; Gasperetti *et al.*, "The Pandemic Will Not Stop Us: The Impact of COVID-19 on Women's Peace Activism in Colombia, the Philippines, South Sudan and Ukraine"; Alexis Henshaw, Digital Frontiers in Gender and Security (Bristol, UK: Bristol University Press, 2023).
25 Gasperetti *et al.*, "The Pandemic Will Not Stop Us: The Impact of COVID-19 on Women's Peace Activism in Colombia, the Philippines, South Sudan and Ukraine."
26 Gasperetti *et al.*

Protection

Promoting the participation of women in online conversations goes hand-in-hand with protecting them. As alluded to above, online harassment is both pervasive and gendered. Approximately 20-30 per cent of all women using social media in developed countries report experiencing gender-based online harassment; that number rises to nearly three-quarters of female users in parts of the Global South.[27] Online abuse is intersectional, with women of color, disabled women, and members of the LGBTQI+ community reporting significantly higher rates of online abuse.[28] Even women in elite positions are not immune to the epidemic of online harassment; over 40 per cent of women parliamentarians and office-holders who are online reporting that they have been targeted with harassment, threats, and/or disinformation.[29]

While militaries cannot (and, arguably, should not) be expected to become content moderators, there are areas in which online violence against women intersects with cybersecurity operations. In particular, as noted above, multiple countries have prioritized the free access and use of cyberspace in their national cyber strategies. Some countries, like the United States, have also noted the importance of upholding international law (including, presumably, human rights protections) online. Threats of gender-based violence online speak directly to these goals. Of particular interest may be the rise of gender-based extremism and the targeting

27 Amnesty International, "Toxic Twitter – A Toxic Place for Women" (London: Amnesty International, 2018), https://www.amnesty.org/en/latest/research/2018/03/online-violence-against-women-chapter-1/; Nithya Sambasivan *et al.*, "'They Don't Leave Us Alone Anywhere We Go': Gender and Digital Abuse in South Asia" (CHI Conference on Human Factors in Computing Systems Proceedings, Glasgow, Scotland, UK, 2019), https://www.classes.cs.uchicago.edu/archive/2020/winter/20370-1/readings/GenderSouthAsia.pdf; UN Women, "Online and ICT* Facilitated Violence against Women and Girls during COVID-19" (New York: UN Women, 2020), https://www.unwomen.org/-/media/headquarters/attachments/sections/library/publications/2020/brief-online-and-ict-facilitated-violence-against-women-and-girls-during-covid-19-en.pdf?la=en&vs=2519.

28 Amnesty International, "Toxic Twitter – A Toxic Place for Women"; Nina Jankowicz *et al.*, "Malign Creativity: How Gender, Sex, and Lies Are Weaponized Against Women Online" (Washington D.C.: The Wilson Center, 2021), https://www.wilsoncenter.org/publication/malign-creativity-how-gender-sex-and-lies-are-weaponized-against-women-online.

29 Lucina Di Meco, "#ShePersisted: Women, Politics, and Power in the New Media World" (Washington D.C.: The Wilson Center, 2019), https://static1.squarespace.com/static/5dba105f102367021c44b63f/t/5dc431aac6bd4e7913c45f7d/1573138953986/191106+SHEPERSISTED_Final.pdf; IPU, "Sexism, Harassment and Violence against Women Parliamentarians," Issues Brief (Geneva, Switzerland: Inter-Parliamentary Union, 2016), http://archive.ipu.org/pdf/publications/issuesbrief-e.pdf.

of women public figures via ICT platforms. Regarding gender-based extremism, researchers have noted an uptick in violence among self-identified "involuntary celibate" or "incel" communities in recent years.[30] Online incel communities, which began as a non-violent and gender-neutral space, have become havens for violent misogyny, disinformation, and conspiracies against women, feminists, and others. Attacks carried out by figures linked to incel communities had killed nearly 50 people in North America by 2020.[31] Incel communities have also been linked to a 2019 attack in Germany and a 2021 mass shooting in the United Kingdom.[32]

While many incel-inspired attacks are carried out by lone actors, online communities provide an important source of radicalization. Research has identified several defining aspects of incel ideology, including beliefs about natural male supremacy, a belief in a feminist conspiracy to undermine government and/or society and a belief that men are entitled to sexual access to women.[33] While incel communities can be racially diverse, incel discourse is often racialized. White supremacist, anti-immigrant, and anti-Semitic speech can be common in these spaces.[34] Though designated incel communities have been banned by some social media platforms, discussions about extremist or violent misogyny remain easy to find. Researchers examining the "manosphere" – the universe of blogs, podcasts, social media channels, and other websites catering to men only – have noted that

30 Bruce Hoffman and Jacob Ware, "Incels: America's Newest Domestic Terrorism Threat," Lawfare, January 12, 2020, https://www.lawfareblog.com/incels-americas-newest-domestic-terrorism-threat; Bruce Hoffman, Jacob Ware, and Ezra Shapiro, "Assessing the Threat of Incel Violence," *Studies in Conflict & Terrorism* 43, no. 7 (July 2, 2020): 565–87, https://doi.org/10.1080/1057610X.2020.1751459.

31 Hoffman and Ware, "Incels."

32 Griffin, "Incels: Inside a Dark World of Online Hate," *BBC News*, August 13, 2021, sec. BBC Trending, https://www.bbc.com/news/blogs-trending-44053828; Daniel Koehler, "The Halle, Germany, Synagogue Attach and the Evolution of the Far-Right Terror Threat," *CTC Sentinel* 12, no. 11 (2019): 14–20.

33 Alex DiBranco, "Male Supremacist Terrorism as a Rising Threat," February 10, 2020, https://icct.nl/publication/male-supremacist-terrorism-as-a-rising-threat/; Roberta Liggett O'Malley, Karen Holt, and Thomas J. Holt, "An Exploration of the Involuntary Celibate (Incel) Subculture Online," *Journal of Interpersonal Violence*, September 24, 2020, 0886260520959625, https://doi.org/10.1177/0886260520959625.

34 DiBranco, "Male Supremacist Terrorism as a Rising Threat"; Ashley Mattheis and Charlie Winter, "'The Greatness of Her Position': Comparing Identitarian and Jihadi Discourses" (London, UK: International Centre for the Study of Radicalisation, 2019), https://icsr.info/wp-content/uploads/2019/05/ICSR-Report-%E2%80%98The-Greatness-of-Her-Position%E2%80%99-Comparing-Identitarian-and-Jihadi-Discourses-on-Women.pdf; Moonshot, "Understanding and Preventing Incel Violence in Canada" (Moonshot CVE, 2021), https://moonshotteam.com/preventing-incel-violence-in-canada/.

these spaces often serve to guide men toward progressively more extreme content.[35]

Dealing with technology's use in violent extremism is one potential concern for military operations. Countering the cyber-harassment of women activists, journalists, and politicians is another. As noted above, countries are inconsistent in their treatment of gender in WPS National Action Plans and Cyber Strategies. The UK's 2022 Cyber Strategy specifically calls for attention to gender-based attacks online and for efforts to ensure that the Internet remains a space for free expression.[36] The U.S. Department of Defense's 2018 Cyber Strategy does not mention "gender" or "women" at all, but it calls for attention to efforts by rogue states to undermine democratic institutions. It also calls for the US to uphold norms of "responsible State behavior in cyberspace."[37]

The targeting of women public figures has likewise been well documented. At least some of this abuse comes from state-sponsored entities in countries like China, Iran, and Russia.[38] Specific examples include the targeting of American scholar and journalist Leta Hong Fincher by Chinese media figures, after she shared information about human rights abuses in the Xinjiang region in 2020, and the cyber-harassment of New York Times cybersecurity reporter Nicole Perlroth by Russian state media outlets the same year.[39] Observers have also noted the spread of disinformation and conspiracy theories against women candidates for public office, which threatens election integrity. Jankowicz *et al.* found intersectional trends in online abuse and conspiracy theories, with women politicians of color in the US among the most frequent targets of gendered disinformation.[40] Much of this also included homophobic or transphobic elements. Common

---

35 See, e.g., Kelly C. Fitzgerald, "Mapping the Manosphere: A Social Network Analysis of the Manosphere on Reddit" (Master's Thesis, Monterey, CA, Naval Postgraduate School, 2020); Jie Liang Lin, "Antifeminism Online: MGTOW (Men GOing Their Own Way)," in *Digital Environments: Ethnographic Perspectives Across Global Online and Offline Spaces*, ed. Urte Undine Frömming *et al.* (Transcript Verlag, 2017). on sub-communities within the manosphere.

36 HM Government, "National Cyber Strategy 2022."

37 Department of Defense, "Department of Defense Cyber Strategy" (Washington DC: Department of Defense, 2018), https://media.defense.gov/2018/Sep/18/2002041658/-1/-1/1/CYBER_STRATEGY_SUMMARY_FINAL.PDF.

38 Di Meco, "#ShePersisted: Women, Politics, and Power in the New Media World"; Jankowicz *et al.*, "Malign Creativity: How Gender, Sex, and Lies Are Weaponized Against Women Online."

39 Jankowicz *et al.*, "Malign Creativity: How Gender, Sex, and Lies Are Weaponized Against Women Online."

40 Jankowicz *et al.*, "Malign Creativity: How Gender, Sex, and Lies Are Weaponized Against Women Online."

themes included claims that the women were transgender, that they had lied about their race and/or country of origin and that they had advanced by cultivating sexual favors. These findings track with similar research from the UK, suggesting that women of color politicians are disproportionately subject to online abuse.[41] Efforts to uphold election integrity and democratic values would certainly call upon cybersecurity officials to take note of these trends – and to try to prevent them.

Prevention

Prevention has been referred to as a marginalized pillar of the WPS agenda, with commentators suggesting that institutionalization frameworks have viewed prevention in reductive ways.[42] In particular, histories of the WPS agenda point out that the civil society advocates and states of the Global South who first lobbied for UNSCR 1325 envisioned the "prevention" pillar as a tool to challenge militarism and even to lobby for disarmament.[43] In practice, however, we have seen prevention largely instrumentalized in the service of specific state or military goals, including aims associated with the War on Terror or countering violent extremism.[44] This tension between broad and narrow operationalizations of prevention should not be overlooked, as there are applications to the digital realm. On the one hand, mainstreaming gender perspectives can serve instrumental/operational goals associated with preventing/countering violent extremism (P/CVE). On the other, returning to the original intent of WPS architects can lead us to reflect on broader questions about the emerging culture and norms that define the use of cyberweapons and the expansion of the cyberarms trade.

In the P/CVE space, many practitioners and commentators have come to lobby not only for gender mainstreaming, but for a more nuanced

41 Amnesty International, "Toxic Twitter – A Toxic Place for Women."

42 Soumita Basu and Catia C. Confortini, "Weakest 'P' in the 1325 Pod? Realizing Conflict Prevention through Security Council Resolution 1325," *International Studies Perspectives* 18, no. 1 (February 1, 2017): 43–63, https://doi.org/10.1093/isp/ekw004.

43 Basu and Confortini; Laura J. Shepherd, *Narrating the Women, Peace and Security Agenda: Logics of Global Governance*, 1st edition (New York, NY: Oxford University Press, 2021).

44 Ann-Kathrin Rothermel, "Gender in the United Nations' Agenda on Preventing and Countering Violent Extremism," *International Feminist Journal of Politics* 22, no. 5 (October 19, 2020): 720–41, https://doi.org/10.1080/14616742.2020.1827967; Laura J. Shepherd, "Making War Safe for Women? National Action Plans and the Militarisation of the Women, Peace and Security Agenda," *International Political Science Review*, March 8, 2016, https://doi.org/10.1177/0192512116629820.

understanding of gender. Critiques of existing initiatives (especially, efforts to combat Islamic extremism in Western countries) note a heavy reliance on crude stereotyping, wherein young men are viewed as the at-risk population while women – in particular mothers – are seen as the natural peacemakers.[45] Such presumptions ignore the widespread involvement of women in, for example, the Islamic State. The involvement of women in policing, fundraising, recruiting, and propagandizing on behalf of the group defies simple stereotypes of gender.[46] Expanding the referent of P/CVE programs to include far-right extremism also requires that we examine the historical and contemporary importance of women as idealogues in these movements.[47] Ebner and Davey point out that putting female voices at the center of far-right online propaganda (including Identitarian and white supremacist movements) has allowed extremist groups to "capitalize on victimhood narratives" that center white women, bolstering their claims that non-white and immigrant populations threaten the safety and sanctity of women.[48] More recently, we have also seen the large-scale mobilization of women into online conspiracy-based movements like QAnon, where the alleged need to protect children – from

45 Phoebe Donnelly, "Demystifying Gender Analysis for Research on Violent Extremism," Researching Violent Extremism Series (RESOLVE Network, 2021), https://www.resolvenet.org/system/files/2021-01/RSVE_RVE%20Series_Donnelly_January%202021.pdf; Rothermel, "Gender in the United Nations' Agenda on Preventing and Countering Violent Extremism"; Emily Winterbotham, "Do Mothers Know Best? How Assumptions Harm CVE," Institute for Global Change, 2018, https://institute.global/policy/do-mothers-know-best-how-assumptions-harm-cve.

46 Carolyn Hoyle, Alexandra Bradford, and Ross Frenett, *Becoming Mulan?: Female Western Migrants to ISIS* (Institute for Strategic Dialogue, 2015); Meredith Loken and Anna Zelenz, "Explaining Extremism: Western Women in Daesh," *European Journal of International Security* 3, no. 1 (February 2018): 45–68, https://doi.org/10.1017/eis.2017.13; Azadeh Moaveni, *Guest House for Young Widows: Among the Women of ISIS* (New York: Random House, 2019); Jessica Trisko Darden, Alexis Henshaw, and Ora Szekely, *Insurgent Women: Female Combatants in Civil Wars*, 1 edition (Washington, DC: Georgetown University Press, 2019).

47 Kathleen M. Blee, *Women of the Klan: Racism and Gender in the 1920s* (University of California Press, 2008); Seyward Darby, *Sisters in Hate: American Women on the Front Lines of White Nationalism*, 2020; Paige Whaley Eager, *From Freedom Fighters to Terrorists* (Ashgate, 2008); Julia Ebner and Jacob Davey, "How Women Advance the Internationalization of the Far-Right" (Washington D.C.: GWU Program on Extremism, 2019), https://extremism.gwu.edu/sites/g/files/zaxdzs2191/f/How%20Women%20Advance%20the%20Internationalization%20of%20the%20Far-Right.pdf.

48 Ebner and Davey, "How Women Advance the Internationalization of the Far-Right."

vaccines, pedophilia, etc. – has justified acts of violence.[49] That much of the misinformation, disinformation, and/or propaganda supporting these movements circulates online demonstrates a need for more inclusive approaches to understanding the pathways by which women as well as men can be radicalized.

At the same time, efforts to prevent violent extremism must grapple with the rise of gender-based extremism and the role of masculinities. Researchers have warned that, in the wake of more aggressive content moderation by major social media platforms, we are seeing the rise of "big tent" conspiracy communities.[50] Facilitated by a mass-movement of extremist voices to less closely moderated social media platforms and encrypted messaging apps, these spaces have become sites for the interchange of various extremist beliefs (including racist, anti-immigrant, and white supremacist views) and conspiracies (including QAnon and general conspiracies about global finance, vaccines, and electoral politics).[51] The influence of extreme voices extends into the "manosphere," the collection of blogs, vlogs, message boards, and social media sites that purport to represent men's interests. Researchers have asserted that the manosphere can become a conduit for far-right radicalization, with both recommendation algorithms and individual users steering conversation from ostensibly more "moderate" communities toward more extreme and/or violent content.[52] Understanding why men and boys are drawn to these online spaces involves reflecting on links between violence and expressions

49 Amarnath Amarasingam and Marc-André Argentino, "The QAnon Conspiracy Theory: A Security Threat in the Making?," Combating Terrorism Center at West Point, July 31, 2020, https://ctc.usma.edu/the-qanon-conspiracy-theory-a-security-threat-in-the-making/; Eden Gillespie, "'Pastel QAnon': The Female Lifestyle Bloggers and Influencers Spreading Conspiracy Theories through Instagram," The Feed, 2020, https://www.sbs.com.au/news/the-feed/pastel-qanon-the-female-lifestyle-bloggers-and-influencers-spreading-conspiracy-theories-through-instagram.

50 Marc-André Argentino *et al.*, "Far From Gone: The Evolution of Extremism in the First 100 Days of the Biden Administration" (London, UK: International Centre for the Study of Radicalisation, 2021), https://icsr.info/wp-content/uploads/2021/04/ICSR-Report-Far-From-Gone-The-Evolution-of-Extremism-in-the-First-100-Days-of-the-Biden-Administration.pdf.

51 Argentino *et al.*; Jacob Davey and Julia Ebner, "The Fringe Insurgency--Connectivity, Convergence and Mainstreaming of the Extreme Right" (London: Institute for Strategic Dialogue, 2017), https://www.isdglobal.org/wp-content/uploads/2017/10/The-Fringe-Insurgency-221017_2.pdf.

52 Colin Clarke and Liana Turner, "The 'Incel' Ideology Continues to Build a Strong Following in the Online 'Manosphere,'" GNET (blog), April 22, 2020, https://gnet-research.org/2020/04/22/the-incel-ideology-continues-to-build-a-strong-following-in-the-online-manosphere/; Manoel Horta Ribeiro *et al.*, "The Evolution of the Manosphere Across the Web," 2021, https://arxiv.org/pdf/2001.07600.pdf; Fitzgerald, "Mapping the Manosphere: A Social Network Analysis of the Manosphere on Reddit."

of masculinity. However, some WPS-inspired interventions in P/CVE have failed to pursue these linkages in a meaningful way.[53] Seeing gender in a more holistic and critical way is arguably essential to successful P/CVE interventions – both online and offline.

Aside from these operational definitions of prevention, it is worth revisiting the origin stories of the WPS. While architects of the agenda initially hoped to see it as a transformative tool, the use of the UN Security Council as a conduit for advancing the agenda meant that some of the more transformative hopes of advocates have not fully been realized.[54] Emerging concerns about cyberwar and cyberweapons, however, may cause us to reflect on these discussions. Conversations about cybersecurity that center systems (not people) as the referent of security, or which abstract from conversations about the potential impact of cyberweapons on humans, should be avoided. Such interventions are incomplete and risk replicating the problematic discussions of earlier eras, for example the dehumanization of nuclear weapons discourse.[55] Perlroth[56] already describes the cyber arms market as a masculinized space in which profit motivations dominate, with little regard for ethical use or human impact. Because cybersecurity is a male-dominated field, professionals may overlook how gender and other social relations shape online threat.[57] Shifting the focus of cybersecurity from people to systems would represent a paradigm shift, but it may be a necessary one given the intertwined nature of people and systems. The examples above clearly demonstrate that offline trends in violence against marginalized groups, including women, have migrated to the online space. These issues have not received sufficient attention from entities engaged in

53 Donnelly, "Demystifying Gender Analysis for Research on Violent Extremism"; David Duriesmith, "Engaging or Changing Men? Understandings of Masculinity and Change in the New 'Men, Peace and Security' Agenda," *Peacebuilding* 0, no. 0 (November 4, 2019): 1–14, https://doi.org/10.1080/21647259.2019.1687076; Hannah Wright, "'Masculinities Perspectives': Advancing a Radical Women, Peace, and Security Agenda?," *International Feminist Journal of Politics* 22, no. 5 (2020): 652–74.

54 Shepherd, *Narrating the Women, Peace and Security Agenda.*

55 Carol Cohn, "Sex and Death in the Rational World of Defense Intellectuals," *Signs* 12, no. 4 (1987): 687–718.

56 Nicole Perlroth, This Is How They Tell Me the World Ends: The Cyberweapons Arms Race, 1st edition (Bloomsbury Publishing, 2021).

57 Nir Kshetri, "The Lack of Women in Cybersecurity Leaves the Online World at Greater Risk," The Conversation, 2020, http://theconversation.com/the-lack-of-women-in-cybersecurity-leaves-the-online-world-at-greater-risk-136654; Winifred R. Poster, "Cybersecurity Needs Women," *Nature* 555, no. 7698 (March 2018): 577–80, https://doi.org/10.1038/d41586-018-03327-w.

cybersecurity. Deeper reflections about who we are securing and how may be necessary to advance cybersecurity discourse.

Relief and Recovery

Concerns about technology and cybersecurity also relate to the relief and recovery pillar of the WPS. For the military audience, these issues might be most closely related to peacekeeping and stability operations. In recent years, we have seen increased attention and scrutiny devoted to the growth of financial technologies and the deployment of technology in support of humanitarian operations. Identity verification and aid disbursement have been presented as ideal use cases for technologies including biometric systems, facial recognition technologies, and blockchain – as discussed below. However, one need look no further than recent conflicts to see where the deployment of such systems can go terribly wrong.

The US military in Afghanistan was engaged in a massive, years-long project to collect biometric data on millions of Afghan citizens, only to see equipment and databases fall into the hands of the Taliban upon their return to power in 2021. The potential risk these data posed, particularly for Afghans who may have assisted US forces, was widely discussed.[58] At the same time, commentators questioned the harms already wrought by aggressive attempts to achieve "identity dominance" through data harvesting. Privacy International alleges that stolen biometric data and devices were being used by the Taliban in local areas that they controlled as early as 2016-2017, a fact which should have prompted greater reflection on the collection and maintenance of this data.[59] Jacobsen further comments on the poor quality of data, resulting from insufficient operator training

58 Eileen Guo and Hikmat Noori, "This Is the Real Story of the Afghan Biometric Databases Abandoned to the Taliban," MIT Technology Review, 2021, https://www.technologyreview.com/2021/08/30/1033941/afghanistan-biometric-databases-us-military-40-data-points/; Ken Klippenstein and Sara Sirota, "The Taliban Have Seized U.S. Military Biometrics Devices," *The Intercept* (blog), August 17, 2021, https://theintercept.com/2021/08/17/afghanistan-taliban-military-biometrics/.

59 Privacy International. "Biometrics and Counter-Terrorism: Case Study of Iraq and Afghanistan." London, UK: Privacy International, 2021. https://privacyinternational.org/sites/default/files/2021-06/Biometrics%20for%20Counter-Terrorism-%20Case%20study%20of%20the%20U.S.%20military%20in%20Iraq%20and%20Afghanistan%20-%20Nina%20Toft%20Djanegara%20-%20v6.pdf.

and / or use of biometric machines in field contexts.[60] She alleges that poorly handled and / or poor-quality data caused misidentifications that may have compromised military justice proceedings against US soldiers accused of war crimes. Even guides developed for US military personnel note that there was no accepted doctrine around biometric collection. However, documentation nonetheless urged military officers to create a culture of support around biometric data collection and use, calling these technologies "almost foolproof."[61] Such sanguine assessments stand at odds with work by researchers and experts, who note the lack of industry standards or independent testing to support the claims of companies developing these technologies.[62]

A gender mainstreamed approach to understanding technology deployment in fragile states and post-conflict circumstances would reflect on how gender and other social hierarchies – including but not limited to race / ethnicity, religion, age, disability, and sexuality or gender identity – shape risk. Given the awareness of potential harms that may result from data overcollection by state, corporate, and / or military actors, such an approach would likely only support data harvesting efforts where they were backed by rigorous privacy policies. These concerns apply equally to the use of technology in pursuit of economic and humanitarian goals.

A demonstration of risk and potential reward can be found in the UN's use of emerging technologies in support of humanitarian assistance programs. Blockchain-based experiments by the World Food Programme (WFP) and UN Women share the goal of using new technologies to establish and safeguard identity. In refugee camps in Jordan, Lebanon, Bangladesh, and Pakistan, the Building Blocks program was established with the goal of providing secure identity verification for displaced persons who lacked other forms of identification (ID). Combining biometric data (iris scans) with a blockchain-based network, the goal was to facilitate

60 Jacobsen, Annie. First Platoon: A Story of Modern War in the Age of Identity Dominance. Penguin, 2021.

61 U.S. Army / CALL, "Commander's Guide to Biometrics in Afghanistan" (Center for Army Lessons Learned (CALL), April 2011), https: / / info.publicintelligence.net / CALL-AfghanBiometrics.pdf.

62 Ali Karimi, "Surveillance in Weak States: The Problem of Population Information in Afghanistan," *International Journal of Communication* 2019, no. 13 (2019): 4778–93; Joseph Pugliese, *Biometrics: Bodies, Technologies, Biopolitics*, 1 online resource (193 pages). vols., Routledge Studies in Science, Technology and Society (London: Taylor and Francis, 2012), http: / / www.123library.org / book_details / ?id=59542; Siraj Shaikh and Christos Dimitriadia, "My Fingers Are All Mine: Five Reasons Why Using Biometrics May Not Be a Good Idea," in *2008 International Symposium on Biometrics and Security Technologies* (IEEE, 2008).

electronic disbursements of aid payments while eliminating the potential for fraud. The use of blockchain, a distributed ledger technology in which encoded data is shared and reconciled across nodes in a network, is meant to safeguard against tampering and to ensure that a network is resilient to the destruction or seizure of its individual nodes. Insofar as data on the blockchain is stored using cryptographic techniques, blockchain advocates argue this technology also offers advances in terms of privacy and control.[63]

While this example may evoke many of the same concerns that surround US biometric programs in Afghanistan, a few key differences can be noted. First, the UN program was carefully piloted before being scaled up. Second, established civil society actors were engaged in project execution and assessment. Third, monitoring and evaluation frameworks mainstreamed a gender perspective through the involvement of UN Women in program development and implementation (this was especially true in Jordan, where the pilot project had a component aimed especially at women). Fourth, program assessments were also conducted by independent experts. Finally, both UN Women and the WFP issued public reports at the conclusion of the pilot program, offering some transparency and public accountability.[64]

Impact studies showed a number of positive benefits to the program. At a practical level, the use of blockchain eliminated the need to share data with or pay transfer fees to financial service providers as part of the process.[65] Women who participated in the program indicated that they felt the system was safer and gave them more control over their funds.[66] Compared to the previous system of cash transfer, in which women would receive disbursements via envelopes with physical cash, the blockchain-based system was viewed as lending more dignity or normalcy to the

63 Victoria L. Lemieux, "Blockchain and Public Record Keeping: Of Temples, Prisons, and the (Re)Configuration of Power," *Frontiers in Blockchain* 2 (2019), https://www.frontiersin.org/article/10.3389/fbloc.2019.00005.

64 Farah Awan and Soheib Nunhuck, "Governing Blocks: Building Interagency Consensus to Coordinate Humanitarian Aid," *Journal of Science Policy and Governance* 16, no. 2 (2020), https://www.sciencepolicyjournal.org/uploads/5/4/3/4/5434385/awan_nunhuck_jspg_v16.2.pdf; UN Women Jordan, "UN Women-WFP Blockchain Pilot Project for Cash Transfers in Refugee Camps" (New York: UN Women, 2021), https://jordan.unwomen.org/en/digital-library/publications/2021/un-women-wfp-blockchain-pilot-project-for-cash-transfers-in-refugee-camps-jordan-case-study; WFP, "Building Blocks | WFP Innovation," World Food Programme, May 22, 2020, https://innovation.wfp.org/project/building-blocks.

65 Awan and Nunhuck, "Governing Blocks: Building Interagency Consensus to Coordinate Humanitarian Aid."

66 UN Women Jordan, "UN Women-WFP Blockchain Pilot Project for Cash Transfers in Refugee Camps."

process.[67] UN Women also engaged in digital literacy outreach as part of the pilot.[68] Overall, this example shows that emerging technologies can effectively support relief and recovery operations, even in adverse contexts and / or where women's initial level of digital literacy may be low.

Still, despite the beneficial results, innovators who wish to integrate emerging technology into relief and recovery efforts face a number of hurdles. First among these are questions of informed consent. While UN Women worked to incorporate data and financial literacy training into the Building Blocks pilot, this training seems to have focused mostly on the use of mobile phones to access and use funds.[69] Deeper conversations about data privacy, the nature of blockchain, and other more technical matters do not appear to have been a part of this program. Outside assessments of Building Blocks and similar initiatives note the problem of achieving informed consent from populations that, for a number of reasons, may be less technologically literate.[70] A particular concern in the space of relief and recovery is that vulnerable communities in the Global South, like refugee camps or impoverished communities, can become laboratories for risky interventions involving financial technologies.[71] This creates important ethical dilemmas. In the case of the Building Blocks program, Awan and Nunhuck further note that Jordan lacks robust data protection laws, meaning that no legal safeguards outside of the UN exist as a recourse if data is mishandled or misused.[72] Similarly, the nature of blockchain technology leaves other questions unanswered. Since records on the blockchain are permanently part of the blockchain, there is no right to be forgotten. It is unclear what happens to data when the humanitarian operations end, or what recourse there is for someone who believes their data is incorrect

67 Awan and Nunhuck, "Governing Blocks: Building Interagency Consensus to Coordinate Humanitarian Aid"; UN Women Jordan, "UN Women-WFP Blockchain Pilot Project for Cash Transfers in Refugee Camps."

68 UN Women Jordan, "UN Women-WFP Blockchain Pilot Project for Cash Transfers in Refugee Camps."

69 UN Women Jordan.

70 Awan and Nunhuck, "Governing Blocks: Building Interagency Consensus to Coordinate Humanitarian Aid"; Margie Cheesman, "Self-Sovereignty for Refugees? The Contested Horizons of Digital Identity," *Geopolitics* 27, no. 1 (2022): 134–59.

71 Cheesman, "Self-Sovereignty for Refugees? The Contested Horizons of Digital Identity."

72 Awan and Nunhuck, "Governing Blocks: Building Interagency Consensus to Coordinate Humanitarian Aid."

or has been misappropriated.[73] Thus, while there is significant potential for emerging technologies in the space of relief and recovery operations, a number of practical and ethical challenges remain.

Toward Institutionalization

This chapter has offered an overview of pathways by which the WPS may be extended into the cyber domain. Examining the pillars of the WPS in turn, we see how emerging issues in the digital space intersect with aims in each area. Each discussion further highlights the possibilities and pitfalls associated with mainstreaming gender in the military application of technology.

This returns to the topic of institutionalization. How can states bring the WPS agenda into the digital realm, without doing so in a reductive and instrumentalizing way? First, as noted earlier, there is much more work that could be done to integrate cyber issues into National Action Plans on WPS. Insofar as NAPs are meant to be synergistic in nature, developed by states in consultation with civil society, they may form an ideal site by which women's groups, data privacy advocates, and other interested actors could discuss gender issues in cyberspace. Relatedly, as governments look to develop cohesive national plans on cybersecurity, consultation with civil society should be prioritized. While "big picture" policies like these may not always impact the day-to-day work of governments, efforts to incorporate the language of gender mainstreaming and to develop associated targets and goals could form the basis for normative change. Eventually, the goal would be to create accountability structures and to ensure that attention to gender cascades through the myriad security agencies responsible for cybersecurity.

Another step toward institutionalization might come in the form of promoting more inclusive cybersecurity. Cultural change in cybersecurity

---

73 Disability can have an effect on biometric identification. In the case of iris scans, medical research has found that certain surgical procedures, especially cataract surgery, can alter the iris in ways that render them incompatible with pre-surgical scans. Roberto Roizenblatt *et al.*, "Iris Recognition as a Biometric Method after Cataract Surgery," *Biomedical Engineering Online* 3 (January 28, 2004): 2, https://doi.org/10.1186/1475-925X-3-2.

will only come about when the sector is diversified.[74] In this way, institutionalization is linked to participation. Observers have suggested many ways to help diversify the profession. Calling attention to gender issues in cybersecurity training and education may be one pathway. Instructors can easily highlight women's many contributions to the field of computing, work to dispel gender stereotypes, or look to include more work by women on their syllabi. In the United States, government agencies have helped to develop cybersecurity education programs at Historically Black Colleges and Universities. Similar programs could be developed at women's colleges or all-girl's schools in other countries. Addressing gender pay gaps and providing re-training opportunities could further help diversify the field. Officials from the U.S. DoD indicate that they have had some success recruiting women veterans to the cybersecurity profession.[75] Because many women leave technology sector jobs in the early career stage, re-training support to help women catch up on skills or make lateral moves from related professions could be beneficial.

Conclusion

This chapter has sought to demonstrate the applicability of the WPS agenda in the cyber domain. Looking at each pillar of the WPS in turn, it shows the range of gendered challenges to cybersecurity. In terms of participation, the movement toward digitizing political debate and the underrepresentation of women in cybersecurity each stand out as challenges to be confronted. In the realm of participation and prevention, there is a clear need to do more to prevent and respond to pervasive gender-based harassment and gendered extremism. In relief and recovery efforts there is potential to use technology in more inclusive ways, but also a need to reflect on the ethical challenge of implementation among vulnerable populations like refugees. Finally, in the discussion of institutionalization there is a call to recognize the potential for

74 It should be noted that this concern applies equally to the need for greater ethnic diversity in the workforce. In the United States, Black, Hispanic, and Native American communities are also underrepresented in cybersecurity. Aspen Institute, "Diversity, Equity, and Inclusion in Cybersecurity" (Washington DC: Aspen Institute, 2021), https://www.aspeninstitute.org/wp-content/uploads/2021/09/Diversity-Equity-and-Inclusion-in-Cybersecurity_9.921.pdf.
75 Terri Moon Cronk, "Women, Minorities Underrepresented in Cybersecurity, DOD Expert Says," U.S. Department of Defense, September 2, 2020, https://www.defense.gov/News/News-Stories/Article/Article/2334909/women-minorities-underrepresented-in-cybersecurity-dod-expert-says/.

dialogue between the WPS Agenda, National Action Plans, and national strategies on cybersecurity. At the same time, there is call for a change in the culture of cybersecurity, urging greater inclusivity.

More broadly, the discussion of WPS in the cyber domain may lead us to think about the need for a paradigm shift in our conceptualization of cybersecurity. Just as the events of the immediate post-Cold War era prompted academics and practitioners to reflect on the referent of security policy (leading to the development of work around human security), the events of our time – including cybercrime, cyber warfare, hybrid warfare, and online violence – may prompt a re-examination of what cybersecurity means. Human Security as a concept was intended to position people and communities, not merely states, as the heart of security policy. This move, while not exclusively feminist in nature, was very much in line with feminist ideas. Cohn and others saw security in the age of weapons of mass destruction as too abstracted from the realities of human risk and suffering.[76] The same might be said of today's cybersecurity apparatus, which focuses on the security of systems while overlooking both the social causes and the human impacts of digital insecurities. When cyberweapons and spyware tools developed for counterterrorism purposes are deployed against civil society activists including feminists,[77] more reflection on ethics and power is clearly called for. The WPS agenda could serve as a tool to empower civil society groups as a check on potentially harmful public-private agreements on cybersecurity and cyberweapons. While addressing the gendered impact of new technologies via WPS is clearly just one part of what should be a larger conversation about ethics, privacy, and policy, conversations about WPS could provide a much-needed opportunity to diversify cybersecurity discussions and provide accountability.

---

76 Cohn, "Sex and Death in the Rational World of Defense Intellectuals."

77 An investigation published in 2021 showed that Pegasus spyware, a tool developed for use in counterterrorism operations, was widely abused by state governments. Pegasus and related tools have been used to target feminist activists and women journalists in India, Mexico, Saudi Arabia, and the United Arab Emirates. See, inter alia, Rina Chandran and Maya Gebeily, "ANALYSIS-From Middle East to India, Women 'violated' in Pegasus Hack," *Reuters*, August 11, 2021, sec. Healthcare, https://www.reuters.com/article/tech-women-surveillance-idUSL8N2P91KX.

References

Amarasingam, Amarnath, and Marc-André Argentino. "The QAnon Conspiracy Theory: A Security Threat in the Making?" *Combating Terrorism Center at West Point*, July 31, 2020. https://ctc.usma.edu/the-qanon-conspiracy-theory-a-security-threat-in-the-making/.

Amnesty International. "Toxic Twitter – A Toxic Place for Women." London: Amnesty International, 2018. https://www.amnesty.org/en/latest/research/2018/03/online-violence-against-women-chapter-1/.

Argentino, Marc-André, Blyth Crawford, Florence Keen, and Hannah Rose. "Far From Gone: The Evolution of Extremism in the First 100 Days of the Biden Administration." London, UK: International Centre for the Study of Radicalisation, 2021. https://icsr.info/wp-content/uploads/2021/04/ICSR-Report-Far-From-Gone-The-Evolution-of-Extremism-in-the-First-100-Days-of-the-Biden-Administration.pdf.

Aspen Institute. "Diversity, Equity, and Inclusion in Cybersecurity." Washington DC: Aspen Institute, 2021. https://www.aspeninstitute.org/wp-content/uploads/2021/09/Diversity-Equity-and-Inclusion-in-Cybersecurity_9.921.pdf.

Awan, Farah, and Soheib Nunhuck. "Governing Blocks: Building Interagency Consensus to Coordinate Humanitarian Aid." *Journal of Science Policy and Governance* 16, no. 2 (2020). https://www.sciencepolicyjournal.org/uploads/5/4/3/4/5434385/awan_nunhuck_jspg_v16.2.pdf.

Basu, Soumita, and Catia C. Confortini. "Weakest 'P' in the 1325 Pod? Realizing Conflict Prevention through Security Council Resolution 1325." *International Studies Perspectives* 18, no. 1 (February 1, 2017): 43–63. https://doi.org/10.1093/isp/ekw004.

Blee, Kathleen M. *Women of the Klan: Racism and Gender in the 1920s*. University of California Press, 2008.

Buzatu, Anne-Marie, Agnieszka Fal-Dutra Santos, Dinah Lakehal, Panthea Pourmalek, and Michaela Zelenanska. "Women, Peace, and Security and Human Rights in the Digital Age: Opportunities and Risks to Advance Women's Meaningful Participation and Protect Their Rights." *Policy Brief*. Global Network of Women Peacebuilders/ICT4Peace Foundation, 2021. https://gnwp.org/wp-content/uploads/PolicyBriefGNWP-2021c.pdf.

Chandran, Rina, and Maya Gebeily. "ANALYSIS-From Middle East to India, Women 'violated' in Pegasus Hack." Reuters, August 11, 2021, sec. Healthcare. https://www.reuters.com/article/tech-women-surveillance-idUSL8N2P91KX.

Cheesman, Margie. "Self-Sovereignty for Refugees? The Contested Horizons of Digital Identity." *Geopolitics* 27, no. 1 (2022): 134–59.

Clarke, Colin, and Liana Turner. "The 'Incel' Ideology Continues to Build a Strong Following in the Online 'Manosphere.'" GNET (blog), April 22, 2020. https://gnet-research.org/2020/04/22/the-incel-ideology-continues-to-build-a-strong-following-in-the-online-manosphere/.

Cohn, Carol. "Sex and Death in the Rational World of Defense Intellectuals." *Signs* 12, no. 4 (1987): 687–718.

Cohoon, J. McGrath, Monica Nable, and Paul Boucher. "Conflicted Identities and Sexism in Computing Graduate Programs." In *2011 Frontiers in Education Conference (FIE)*, S1H-1-S1H-5, 2011. https://doi.org/10.1109/FIE.2011.6142915.

Crowther, Glenn Alexander. "The Cyber Domain." *The Cyber Defense Review*, Fall 2017 (2017): 63–78.

Darby, Seyward. *Sisters in Hate: American Women on the Front Lines of White Nationalism*, 2020.

Davey, Jacob, and Julia Ebner. "The Fringe Insurgency--Connectivity, Convergence and Mainstreaming of the Extreme Right." London: Institute for Strategic Dialogue, 2017. https://www.isdglobal.org/wp-content/uploads/2017/10/The-Fringe-Insurgency-221017_2.pdf.

Department of Defense. "Department of Defense Cyber Strategy." Washington DC: Department of Defense, 2018. https://media.defense.gov/2018/Sep/18/2002041658/-1/-1/1/CYBER_STRATEGY_SUMMARY_FINAL.PDF.

Di Meco, Lucina. "#ShePersisted: Women, Politics, and Power in the New Media World." Washington D.C.: The Wilson Center, 2019. https://static1.squarespace.com/static/5dba105f102367021c44b63f/t/5dc431aac6bd4e7913c45f7d/1573138953986/191106+SHEPERSISTED_Final.pdf.

DiBranco, Alex. "Male Supremacist Terrorism as a Rising Threat," February 10, 2020. https://icct.nl/publication/male-supremacist-terrorism-as-a-rising-threat/.

Donnelly, Phoebe. "Demystifying Gender Analysis for Research on Violent Extremism." Researching Violent Extremism Series. RESOLVE Network, 2021. https://www.resolvenet.org/system/files/2021-01/RSVE_RVE%20Series_Donnelly_January%202021.pdf.

Duriesmith, David. "Engaging or Changing Men? Understandings of Masculinity and Change in the New 'Men, Peace and Security' Agenda." *Peacebuilding* 8, no. 4, 418-431. https://doi.org/10.1080/21647259.2019.1687076.

Eager, Paige Whaley. From Freedom Fighters to Terrorists. Ashgate, 2008.

Ebner, Julia, and Jacob Davey. "How Women Advance the Internationalization of the Far-Right." Washington D.C.: GWU Program on Extremism, 2019. https://extremism.gwu.edu/sites/g/files/zaxdzs2191/f/How%20Women%20Advance%20the%20Internationalization%20of%20the%20Far-Right.pdf.

Enloe, Cynthia. *The Big Push: Exposing and Challenging the Persistence of Patriarchy*. Myriad Editions, 2017.

Falkner, Katrina, Claudia Szabo, Dee Mitchell, Anna Szorenyi, and Shantel Thyer. "Gender Gap in Academia: Perceptions of Female Computer Science Academics." *Proceedings of the 2015 ACM Conference on Innovation and Technology in Computer Science Education*, 2015. https://dl.acm.org/doi/pdf/10.1145/2729094.2742595?casa_token=OOQi6kGVSwQAAAAA:U7SDNWSTd6c4mcJMmz6JI6sfuAhb2ynpvXhfTLa-cc7jjFyU1-3TbAdIFiD2QzQQN2ZHQyiH42Sa.

Fitzgerald, Kelly C. "Mapping the Manosphere: A Social Network Analysis of the Manosphere on Reddit." Master's Thesis, Naval Postgraduate School, 2020.

Frost & Sullivan. "The 2017 Global Information Security Workforce Study: Women in Cybersecurity." Clearwater, FL: Center for Cyber Safety and Education/Frost & Sullivan, 2017. https://www.isc2.

org/-/media/Files/Research/ISC2-Women-in-Cybersecurity-2017.ashx?la=en&hash=FA78E4BDA2F858A0D3CFAC75AF9E789369A0BC8D.

Gasperetti, Heidi, Sena Bölükoğlu, Christine Choi, and B. Kathryn Baron. "The Pandemic Will Not Stop Us: The Impact of COVID-19 on Women's Peace Activism in Colombia, the Philippines, South Sudan and Ukraine." *Case Study Reports*. New York, NY: Global Network of Women Peacebuilders/Center for Global Affairs at NYU, March 2021. https://gnwp.org/wp-content/uploads/NYU-Research-Policy-Brief_March-2021.pdf.

Gillespie, Eden. "'Pastel QAnon': The Female Lifestyle Bloggers and Influencers Spreading Conspiracy Theories through Instagram." *The Feed*, 2020. https://www.sbs.com.au/news/the-feed/pastel-qanon-the-female-lifestyle-bloggers-and-influencers-spreading-conspiracy-theories-through-instagram.

Government of Ireland. "Women, Peace and Security: Ireland's Third National Action Plan for the Implementation of UNSCR 1325 and Related Resolutions." Dublin: Government of Ireland, 2019. https://www.dfa.ie/media/dfa/ourrolepolicies/womenpeaceandsecurity/Third-National-Action-Plan.pdf.

Griffin. "Incels: Inside a Dark World of Online Hate." BBC News, August 13, 2021, sec. BBC Trending. https://www.bbc.com/news/blogs-trending-44053828.

Guo, Eileen, and Hikmat Noori. "This Is the Real Story of the Afghan Biometric Databases Abandoned to the Taliban." *MIT Technology Review*, 2021. https://www.technologyreview.com/2021/08/30/1033941/afghanistan-biometric-databases-us-military-40-data-points/.

Hamilton, Caitlin, and Laura J. Shepherd. "WPS National Action Plans: Content Analysis and Data Visualization, V2," 2020. http://www.wpsnaps.org.

Hayden, Michael V. "The Future of Things Cyber." In *Understanding Cyber Security: Emerging Governance and Strategy*, xi–xv. New York, N.Y: Rowman & Littlefield, 2018.

Henshaw, Alexis. *Digital Frontiers in Gender and Security*. Bristol, UK: Bristol University Press, 2023.

Henshaw, Alexis Leanna. "'Peace with a Woman's Face': Women, Social Media, and the Colombian Peace Process." *Contexto Internacional* 42, no. 3 (2020): 515–38.

HM Government. "National Cyber Strategy 2022." London, UK: HM Government, 2022. https://assets.publishing.service.gov.uk/government/uploads/system/uploads/attachment_data/file/1053023/national-cyber-strategy-amend.pdf.

Hoffman, Bruce, and Jacob Ware. "Incels: America's Newest Domestic Terrorism Threat." *Lawfare*, January 12, 2020. https://www.lawfareblog.com/incels-americas-newest-domestic-terrorism-threat.

Hoffman, Bruce, Jacob Ware, and Ezra Shapiro. "Assessing the Threat of Incel Violence." *Studies in Conflict & Terrorism* 43, no. 7 (July 2, 2020): 565–87. https://doi.org/10.1080/1057610X.2020.1751459.

Horta Ribeiro, Manoel, Jeremy Blackburn, Barry Bradlyn, Emiliano De Cristofaro, Gianluca Stringhini, Summer Long, Stephanie Greenberg, and Savvas Zannettou. "The Evolution of the Manosphere Across the Web," 2021. https://arxiv.org/pdf/2001.07600.pdf.

Hoyle, Carolyn, Alexandra Bradford, and Ross Frenett. *Becoming Mulan?: Female Western Migrants to ISIS*. Institute for Strategic Dialogue, 2015.

Human Rights Centre. "Ways to Bridge the Gender Digital Divide from a Human Rights Perspective." Essex, UK: Human Rights, Big Data and Technology Project at the Human Rights Centre of the University of Essex, n.d. https://www.ohchr.org/Documents/Issues/Women/WRGS/GenderDigital/HRBDT_submission.pdf.

IPU. "Sexism, Harassment and Violence against Women Parliamentarians." Issues Brief. Geneva, Switzerland: Inter-Parliamentary Union, 2016. http://archive.ipu.org/pdf/publications/issuesbrief-e.pdf.

(ISC)2. "Women in Cybersecurity: Young, Educated, and Ready to Take Charge." International Information System Security Certification Consortium, Inc., 2019. https://www.isc2.org/-/media/ISC2/Research/ISC2-Women-in-Cybersecurity-Report.ashx?la=en&hash=4C3B33AABFBEAFDDA211856CB274EBDDF9DBEB38.

Jacobsen, Annie. *First Platoon: A Story of Modern War in the Age of Identity Dominance*. Penguin, 2021.

Jankowicz, Nina, Jillian Hunchak, Alexandra Pavliuc, Celia Davies, Shannon Pierson, and Zoë Kaufmann. "Malign Creativity: How Gender, Sex, and Lies Are Weaponized Against Women Online." Washington D.C.: The Wilson Center, 2021. https://www.wilsoncenter.org/publication/malign-creativity-how-gender-sex-and-lies-are-weaponized-against-women-online.

Karimi, Ali. "Surveillance in Weak States: The Problem of Population Information in Afghanistan." *International Journal of Communication* 2019, no. 13 (2019): 4778–93.

Klippenstein, Ken, and Sara Sirota. "The Taliban Have Seized U.S. Military Biometrics Devices." The Intercept (blog), August 17, 2021. https://theintercept.com/2021/08/17/afghanistan-taliban-military-biometrics/.

Koehler, Daniel. "The Halle, Germany, Synagogue Attach and the Evolution of the Far-Right Terror Threat." *CTC Sentinel* 12, no. 11 (2019): 14–20.

Kreuzer, Michael. "Cyberspace Is an Analogy, Not a Domain: Rethinking Domains and Layers of Warfare for the Information Age." *The Strategy Bridge*, July 8, 2021. https://thestrategybridge.org/the-bridge/2021/7/8/cyberspace-is-an-analogy-not-a-domain-rethinking-domains-and-layers-of-warfare-for-the-information-age.

Kshetri, Nir. "The Lack of Women in Cybersecurity Leaves the Online World at Greater Risk." *The Conversation*, 2020. http://theconversation.com/the-lack-of-women-in-cybersecurity-leaves-the-online-world-at-greater-risk-136654.

Kshetri, Nir, and Maya Chhetri. "Gender Asymmetry in Cybersecurity: Socioeconomic Causes and Consequences." *Computer*, no. February 2022 (2022): 72–77.

Lemieux, Victoria L. "Blockchain and Public Record Keeping: Of Temples, Prisons, and the (Re)Configuration of Power." *Frontiers in Blockchain* 2 (2019). https://www.frontiersin.org/article/10.3389/fbloc.2019.00005.

Liang Lin, Jie. "Antifeminism Online: MGTOW (Men GOing Their Own Way)." In *Digital Environments: Ethnographic Perspectives Across Global Online and Offline Spaces*, edited by Urte Undine Frömming, Steffen Köhn, Samantha Fox, and Mike Terry. Transcript Verlag, 2017.

Loken, Meredith, and Anna Zelenz. "Explaining Extremism: Western Women in Daesh." *European Journal of International Security* 3, no. 1 (February 2018): 45–68. https://doi.org/10.1017/eis.2017.13.

Mattheis, Ashley, and Charlie Winter. "'The Greatness of Her Position': Comparing Identitarian and Jihadi Discourses." London, UK: International Centre for the Study of Radicalisation, 2019. https://icsr.info/wp-content/uploads/2019/05/ICSR-Report-%E2%80%98The-Greatness-of-Her-Position%E2%80%99-Comparing-Identitarian-and-Jihadi-Discourses-on-Women.pdf.

Meyer McAleese, Mary K. "WPS and the Organization of American States." In *The Oxford Handbook of Women, Peace, and Security*, edited by Sara E. Davies and Jacqui True, 413–27. Oxford; New York: Oxford University Press, 2019.

Moaveni, Azadeh. *Guest House for Young Widows: Among the Women of ISIS*. New York: Random House, 2019.

Moon Cronk, Terri. "Women, Minorities Underrepresented in Cybersecurity, DOD Expert Says." U.S. Department of Defense, September 2, 2020. https://www.defense.gov/News/News-Stories/Article/Article/2334909/women-minorities-underrepresented-in-cybersecurity-dod-expert-says/.

Moonshot. "Understanding and Preventing Incel Violence in Canada." *Moonshot CVE*, 2021. https://moonshotteam.com/preventing-incel-violence-in-canada/.

Nagarajan, Chitra, and Azadeh Moaveni. "Another Deeply Gendered War Is Being Waged in Ukraine," March 15, 2022. https://www.aljazeera.com/opinions/2022/3/15/another-deeply-gendered-war-is-being-waged-in-ukraine.

NATO. "Cyber Defence Pledge." NATO, July 8, 2016. https://www.nato.int/cps/en/natohq/official_texts_133177.htm.

OECD. "Bridging the Digital Gender Divide: Include, Upskill, Innovate." Paris, France: OECD, 2018. http://www.oecd.org/digital/bridging-the-digital-gender-divide.pdf.

O'Malley, Roberta Liggett, Karen Holt, and Thomas J. Holt. "An Exploration of the Involuntary Celibate (Incel) Subculture Online." *Journal of Interpersonal Violence*, September 24, 2020, 0886260520959625. https://doi.org/10.1177/0886260520959625.

OSESGY. "Cutting-Edge Tech in the Service of Inclusive Peace in Yemen." OSESGY, August 3, 2020. https://osesgy.unmissions.org/cutting-edge-tech-service-inclusive-peace-yemen.

———. "UN Special Envoy's Office for Yemen Concludes First Online Mass Consultations." OSESGY, June 10, 2020. https://osesgy.unmissions.org/un-special-envoy%E2%80%99s-office-yemen-concludes-first-online-mass-consultations.

Perlroth, Nicole. *This Is How They Tell Me the World Ends: The Cyberweapons Arms Race*. 1st edition. Bloomsbury Publishing, 2021.

Poster, Winifred R. "Cybersecurity Needs Women." *Nature* 555, no. 7698 (March 2018): 577–80. https://doi.org/10.1038/d41586-018-03327-w.

Privacy International. "Biometrics and Counter-Terrorism: Case Study of Iraq and Afghanistan." London, UK: Privacy International, 2021. https://privacyinternational.org/sites/default/files/2021-06/Biometrics%20for%20

Counter-Terrorism-%20Case%20study%20of%20the%20U.S.%20military%20in%20Iraq%20and%20Afghanistan%20-%20Nina%20Toft%20Djanegara%2-0-%20v6.pdf.

Pugliese, Joseph. *Biometrics: Bodies, Technologies, Biopolitics.* 1 online resource (193 pages). vols. Routledge Studies in Science, Technology and Society. London: Taylor and Francis, 2012. http://www.123library.org/book_details/?id=59542.

Radu, Camélia, and Nadia Smaili. "Board Gender Diversity and Corporate Response to Cyber Risk: Evidence from Cybersecurity Related Disclosure." *Journal of Business Ethics* 177, no. 2 (May 1, 2022): 351–74. https://doi.org/10.1007/s10551-020-04717-9.

Raymond, Mark. "Managing Decentralized Cyber Governance." In *Understanding Cyber Security: Emerging Governance and Strategy*, 25–48. New York, N.Y: Rowman & Littlefield, 2018.

Republic of Namibia. "Namibia National Action Plan on Women Peace and Security." Windhoek, Namibia: Republic of Namibia, 2019. https://www.peacewomen.org/sites/default/files/Namibia%20NAP%20(2019-2024).pdf.

Roizenblatt, Roberto, Paulo Schor, Fabio Dante, Jaime Roizenblatt, and Rubens Belfort. "Iris Recognition as a Biometric Method after Cataract Surgery." *Biomedical Engineering* Online 3 (January 28, 2004): 2. https://doi.org/10.1186/1475-925X-3-2.

Rothermel, Ann-Kathrin. "Gender in the United Nations' Agenda on Preventing and Countering Violent Extremism." *International Feminist Journal of Politics* 22, no. 5 (October 19, 2020): 720–41. https://doi.org/10.1080/14616742.2020.1827967.

Sambasivan, Nithya, Amna Batool, Nova Ahmed, Tara Matthews, Kurt Thomas, Laura Sanely Gaytán-Lugo, David Mener, Elie Bursztein, Elizabeth Churchill, and Sunny Consolvo. "'They Don't Leave Us Alone Anywhere We Go': Gender and Digital Abuse in South Asia." Glasgow, Scotland, UK, 2019. https://www.classes.cs.uchicago.edu/archive/2020/winter/20370-1/readings/GenderSouthAsia.pdf.

Saujani, Reshma. "Disrupting Tech's Growing Gender Gap." Presented at the Qualtrics/Girls Who Code Forum, Online, January 27, 2021.

SDG Knowledge Hub. "What Is the Gender Digital Divide, and Why Should It Matter for the SDGs?," IISD, 2017. https://sdg.iisd.org:443/commentary/guest-articles/what-is-the-gender-digital-divide-and-why-should-it-matter-for-the-sdgs/.

Shaikh, Siraj, and Christos Dimitriadia. "My Fingers Are All Mine: Five Reasons Why Using Biometrics May Not Be a Good Idea." In *2008 International Symposium on Biometrics and Security Technologies*. IEEE, 2008.

Shepherd, Laura J. "Making War Safe for Women? National Action Plans and the Militarisation of the Women, Peace and Security Agenda." *International Political Science Review*, March 8, 2016. https://doi.org/10.1177/0192512116629820.

———. *Narrating the Women, Peace and Security Agenda: Logics of Global Governance.* 1st edition. New York, NY: Oxford University Press, 2021.

Trisko Darden, Jessica, Alexis Henshaw, and Ora Szekely. *Insurgent Women: Female Combatants in Civil Wars.* 1 edition. Washington, DC: Georgetown University Press, 2019.

UN Women. "Online and ICT* Facilitated Violence against Women and Girls during COVID-19." New York: UN Women, 2020. https://www.unwomen.org/-/media/headquarters/attachments/sections/library/publications/2020/brief-online-and-ict-facilitated-violence-against-women-and-girls-during-covid-19-en.pdf?la=en&vs=2519.

———. "Women's Meaningful Participation in Negotiating Peace and the Implementation of Peace Agreements." New York: UN Women, 2018. https://www.unwomen.org/-/media/headquarters/attachments/sections/library/publications/2018/egm-womens-meaningful-participation-in-negotiating-peace-en.pdf?la=en&vs=3047.

UN Women Jordan. "UN Women-WFP Blockchain Pilot Project for Cash Transfers in Refugee Camps." New York: UN Women, 2021. https://jordan.unwomen.org/en/digital-library/publications/2021/un-women-wfp-blockchain-pilot-project-for-cash-transfers-in-refugee-camps-jordan-case-study.

U.S. Army/CALL. "Commander's Guide to Biometrics in Afghanistan." Center for Army Lessons Learned (CALL), April 2011. https://info.publicintelligence.net/CALL-AfghanBiometrics.pdf.

U.S. Cyber Command. "Mission and Vision." U.S. Cyber Command, n.d. https://www.cybercom.mil/About/Mission-and-Vision/.

WFP. "Building Blocks | WFP Innovation." World Food Programme, May 22, 2020. https://innovation.wfp.org/project/building-blocks.

Winterbotham, Emily. "Do Mothers Know Best? How Assumptions Harm CVE." Institute for Global Change, 2018. https://institute.global/policy/do-mothers-know-best-how-assumptions-harm-cve.

Wright, Hannah. "'Masculinities Perspectives': Advancing a Radical Women, Peace, and Security Agenda?" *International Feminist Journal of Politics* 22, no. 5 (2020): 652–74.

9

EDUCATING SECURITY

The Women, Peace and Security Agenda in Professional Military Education and Training

Max Thompson and Andree-Anne Melancon[1]

Implementing the Women, Peace and Security agenda in military operations requires effective integration into professional military education. Paragraph 6 of UNSCR 1325 states that the UN Secretary General must provide "training guidelines and materials on the protection, rights and the particular needs of women, as well as on the importance of involving women in all peacekeeping and peace-building measures."[2] This training has been a welcome addition in contributing to the effectiveness of military planning, operations, and relief efforts. However, this falls short of setting a requirement for the use or integration of these training guidelines/materials in national curricula in the same way as dissemination requirements have been integrated in other key international frameworks. Article 83 of the First Additional Protocol for the Geneva Conventions states, for example, that the convention must be disseminated in times of peace "and, in particular, to include the study thereof in their programmes of military instruction and to encourage the study thereof by the civilian population, so that those

1 This chapter reflects the work of the entire department of Defence and International Affairs at the Royal Military Academy Sandhurst and is indebted to all members of the department especially Dr Edward Flint, Head of Department.
2 United Nations Security Council, "Security Council Resolution 1325", S/RES/1325, 31 October 2000. S/RES/1325 (2000). http://unscr.com/en/resolutions/doc/1325.

instruments may become known to the armed forces and to the civilian population."[3]

In short, when it comes to the Women, Peace and Security agenda ***training*** is mandated (especially in times of conflict), but ***education*** is not. This distinction is key since it shapes the way WPS is integrated into professional military education (PME). Training focusses on practical concrete applications. It is concerned with doing and relies on models and short-term processes in order to prepare for the tasks ahead. Education focusses on knowledge and the development of reasoning and judgement. It is a longer-term process. Broadly, training is often described as 'learning to do' whereas education is 'learning to know'.

Some attention has been given to the implementation and dissemination of the WPS agenda at the national level. On the second anniversary of the adoption of UNSCR 1325, the Security Council reaffirmed its commitment to its full implementation and requested "the Secretary-General to prepare a follow-up report on the full implementation of resolution 1325 to be presented to the Security Council in October 2004."[4] Following the submission of this report, the President of the Security Council issued a statement on 28 October 2004 introducing the concept of National Action Plans on the implementation of the WPS agenda.[5] This call for the establishment of National Action Plans was reiterated in a statement on 27 October 2005.[6] These have become a key tool to operationalise and integrate the international mandate of the WPS agenda in national frameworks. In June 2005, Denmark became the first country to adopt a WPS National Action Plan which was heavily focused on the 'participation' pillar of UNSCR 1325. As of September 2022, 103 countries (53 per cent of United Nations Members States) have adopted a WPS National Action Plan, of which only 36 per cent include a budget for implementation.[7]

3 International Committee of the Red Cross, "Protocol Additional to the Geneva Conventions of 12 August 1949, and relating to the Protection of Victims of International Armed Conflicts (Protocol I)", 8 June 1977. https://ihl-databases.icrc.org/ihl/INTRO/470.

4 United Nations Security Council, "Statement by the President of the Security Council", S/PRST/2002/32, 31 October 2002. https://www.securitycouncilreport.org/atf/cf/%7B65BFCF9B-6D27-4E9C-8CD3-CF6E4FF96FF9%7D/WPS%20SPRST200232.pdf.

5 United Nations Security Council, "Statement by the President of the Security Council", S/PRST/2004/40, 28 October 2004. https://www.securitycouncilreport.org/atf/cf/%7B65BFCF9B-6D27-4E9C-8CD3-CF6E4FF96FF9%7D/Disarm%20SPRST200440.pdf.

6 United Nations Security Council, "Statement by the President of the Security Council", S/PRST/2005/52, 27 October 2005. https://www.securitycouncilreport.org/atf/cf/%7B65BFCF9B-6D27-4E9C-8CD3-CF6E4FF96FF9%7D/WPS%20SPRST200552.pdf.

7 Gender Action for Peace and Security UK, "1325 National Action Plans (NAPs)". https://1325naps.peacewomen.org.

Strategic outcome 7: UK capabilities	
HMG continues to strengthen its capability, processes and leadership to deliver against WPS commitments	
Indicator	Indicator source
Mainstreaming of gender in National Scurity Council country strategies	New
Integration of women, peace and security into cross-departmental operational plans and programmes at country level	New
Mainstreaming of gender where appropriate and applicable in military doctrine and training materials, and appointment military gender advisers	New
Integration of gender analysis into conflict analysis and research	New
Training and capacity building of HMG staff to implement women, peace and security commitments	New

Table 9.1. UK NAP 2018-2022 Strategic Outcome 7
Source: UK National Action Plan on Women, Peace and Security 2018–2022, 26.

The United Kingdom's (UK) fourth National Action Plan was published in January 2018 and offers several examples of WPS training offered and supported overseas (Defence Engagement) especially in fragile and conflict-affected states (FCAS) but has a relatively minimalist approach to training government forces and civil servants. Shown in Table 9.1, Strategic Outcome 7 focuses on "UK Capabilities" and sets out several indicators.

The Gender Action for Peace and Security (GAPS) provide an annual report: "Assessing UK Government Action on Women, Peace and Security in 2020." It found that:

> COVID-19 has had a severe impact on UK Government WPS training. Whilst there is a WPS training online, the intensive Gender Conflict and Stability has not moved online. GAPS hopes that the Gender, Conflict and Stability training can restart as soon as possible, and that future NAPs [National Action Plans] commit to training all staff who work in, or on, FCAS. Training is essential for NAP implementation and should be provided systemically and comprehensively, including tailored training that is appropriate to roles and responsibilities, and contexts.[8]

8 "Assessing UK Government Action on Women, Peace and Security in 2020", April 2021. https://gaps-uk.org/wp-content/uploads/2021/04/GAPS-Shadow-Report-2020.pdf .

Centralising training is emphasised, and the report goes on to recommend that the UK government should "Ensure the Gender, Conflict and Stability training restarts as soon as possible and that gender is included in all other related training which all Government officials working in, or on, FCAS undertake. Ensure the inclusion of WPS into all UK Government official job descriptions of staff working in, or on, FCAS."[9] This addition has been welcomed as UK defence has sought to integrate WPS into defence doctrine through Joint Services Publication (JSP) 1325 *Human Security in Military Operations* in 2019 and JSP985 *Human Security in Defence* in 2021.[10] Education is, however, largely absent from the UK's approach to WPS. For example, the section titled 'Integrating Human Security into Training' of JSP 985 is only 2.5 pages long. It discusses future actions, such as a central training repository that *will* be designed and provides little detailed information instead opting for generic phrases such as "a through-career HS education and training pathway needs to be implemented."[11]

Much has been written on various aspects of the implementation of the WPS agenda and the aim here is not to present an overview of this literature. Rather, it is to highlight one aspect that is seldom explored and has to do with the integration of WPS in PME. For example, of the ten chapters on implementing WPS in *The Oxford Handbook on Women, Peace, and Security* (2019) just one case, that of the Australia Defence Forces (ADF) explored WPS in the armed forces and only mentions PME when stating: "work has commenced to include WPS into the professional military and training continuum for all ADF personnel, and currently training is conducted for officers attending junior and senior command and staff colleges."[12] Part of this oversight can be explained by the divide between academia and practitioners. As Ambassador Melanne Verveer of the Georgetown Institute for Women, Peace and Security argues: "There are tensions between the academic sector and the people who are doing the practical work that we do in development or diplomacy or with the military. I think those gaps have to be bridged and one of the things we

9 Ibid.

10 See Annex 1 of this book for overview of JSP985.

11 UK Ministry of Defence, "JSP 985: Human Security in Defence" December 2021. https://assets.publishing.service.gov.uk/government/uploads/system/uploads/attachment_data/file/1040257/20211209_JSP_985_Vol_1.pdf.

12 Jennifer Witter, "Mainstreaming WPS in the Armed Forces: The Case of Australia" in *The Oxford Handbook of Women, Peace, and Security*, Eds. Sara E. Davies and Jacqui True (Oxford: Oxford University Press, 2019), 10-11. https://jenniferwittwer.com/wp-content/uploads/2019/01/oxfordhb-WPS-in-Armed-Forces.pdf .

actually set up was a programmatic framework on bridging theory and practice."[13] This is where PME comes in. Indeed, professional military education encompasses both academia and the military and thus, should be at the forefront of the implementation of the WPS agenda.

Drawing on the authors' professional experiences in integrating WPS into a PME establishment at the Royal Military Academy Sandhurst (RMAS), this chapter examines the challenges of integrating the WPS agenda into military training and education. This chapter investigates how PMEs can effectively adapt to address and recognise the importance of the pillars of WPS and gender mainstreaming. It will also explore the unique institutional limitation impeding the implementation of WPS in military education. This chapter argues that effective integration of WPS and human security in PME requires more than a syllabus update ('add WPS and stir'), it necessitates an incremental revamping of assessments, readings, and methods, leading to the implementation of a whole-course approach. It dubs this **the Sandhurst Approach** and offers insights that may be useful for setting out best practices for other military academies and education centres. As such it concludes with a series of recommendations for the effective mainstreaming of WPS within PME.

Professional Military Education

There are a variety of challenges that professional military educators confront due to the unique setting and context in which teaching is delivered, this chapter highlights five and offers insights on how best to integrate WPS in a manner that addresses them. The five challenges are:

1. Time
2. Interest
3. Professional relevance
4. Institutional biases
5. Cultural awareness

The authors are senior lecturers at RMAS. RMAS provides a 44-week commissioning course centred on leadership that takes civilian students,

13 Cited in Erin Cooper, "Teaching the WPS Agenda in a Changing Security Landscape", Our Secure Future Policy Brief, January 2021. https://oursecurefuture.org/sites/default/files/Teaching-WPS-Agenda-Changing-Security-Landscape.pdf.

officer cadets (OCdts), and transforms them into British Army officers capable of leading soldiers and operating in complex environments. This necessarily involves developing their understanding of the international security environment. RMAS has three academic departments: Communication and Applied Behavioural Sciences (CABS), Defence and International Affairs (DIA), and War Studies (WS). These make up the Faculty of Leadership, Security, and Warfare (LSW). OCdts complete mandatory and critical training in these three fields and their education is assessed and appraised through the Army Higher Education Pathway. RMAS recognise that:

> Academic study is an integral part in developing effective and professional military leaders. The courses provided by the Faculty for the study of Leadership, Security, and Warfare (LSW) Departments are purposely designed to develop an officer's understanding and analysis of the human, operational and strategic context in which they will work.[14]

In practice, this means that through a series of practical and academic exercises, senior lecturers educate and assess how OCdts can understand the context in which they operate and analyse an international security environment with evolving threats and strategic trends. DIA have 30 90-minute periods over 44 weeks in which to cover and assess a wide range of academic topics including UK foreign and defence policies, international law (*jus ad bellum/jus in bello*), concepts (power, security, conflict), and the range of international security responses available to states to manage conflict. A full list of sessions covered is found in Table 9.3 later in this chapter. This is covered in two modules "the international security environment" and "international security responses." Though necessarily covering a range of topics, the four core aspects of what DIA cover can be summarised as:

1. Security
2. Conflict Management
3. International Law
4. UK Foreign and Defence Policy

Each of these aspects should involve awareness of the WPS agenda. **Educating Security** requires an awareness of firstly the academic and

14 Army, "RMA Sandhurst: Faculty for the Study of Leadership, Security, and Warfare", https://www.army.mod.uk/who-we-are/our-schools-and-colleges/rma-sandhurst/faculty-for-the-study-of-leadership-security-and-warfare/.

policy shift from nuclear and state-centred security to a broader and deeper understanding of human security, that security affects individuals as well as states, and secondly the WPS agenda, that men, women, boys, and girls experience conflict differently. It also relates to the identification of wider Global Strategic Trends to facilitate strategic thinking and foreign policy planning.[15] **Conflict Management** requires assessing the tools and measures states have at their disposal to end violence and build positive peace,[16] these include: conflict mediation and other forms of preventive diplomacy; stabilisation and peacekeeping; peace-enforcement activities; reforms including disarmament, demobilisation and reintegration (DDR) and Security Sector reform (SSR); and grass roots engagement with civil society and other local groups. This covers all aspects of the WPS pillars. **International Law** covers the general laws of war, but also developing an awareness of international legal regimes covering human trafficking, the protection of children, and rape as a weapon of war. **UK Foreign and Defence Policy** requires education in the UK's human security approach and National Action Plans.

There are five general challenges that a professional military academy has that differ from the university sector. The first issue which students and lecturers confront is ***time.*** Students are necessarily receiving a vast amount of training and the number of hours in their day that can be devoted towards academic study is small compared with what universities can expect. Similarly, the teaching load of academic staff tends to be higher than higher education equivalents. The second issue is ***interest.*** Unlike universities where students have opted-in to study their chosen course, students in most professional military academies are mandated to learn certain topics including security and conflict management regardless of their level of background interest. The mandating of certain topics also affects staff who may have to cover topics they are less familiar with. The third issue, related to the first two, is ***professional relevance***. It is

15 UK Ministry of Defence, "Global Strategic Trends: The Future Starts Today" 6th edition, 2 October 2018. https://assets.publishing.service.gov.uk/government/uploads/system/uploads/attachment_data/file/1075981/GST_the_future_starts_today.pdf.
16 UK Stabilisation Unit, "Joint Analysis of Conflict and Stability (JACS) Guidance Note", 1 June 2017. https://assets.publishing.service.gov.uk/government/uploads/system/uploads/attachment_data/file/765448/JACS_Guidance_Note.pdf.
UK Stabilisation Unit, "The UK Government's approach to Stabilisation: A guide for policy makers and practitioners", 19 December 2018. https://assets.publishing.service.gov.uk/government/uploads/system/uploads/attachment_data/file/784001/The_UK_Government_s_Approach_to_Stabilisation_A_guide_for_policy_makers_and_practitioners.pdf.

particularly important to demonstrate that the material and topics chosen and delivered are relevant for future careers. "Security" and "Conflict management" are not just abstractions for OCdts, but need to be applied. The fourth issue is ***institutional biases*** and the tendency for certain topics which may appear less obviously relevant to military training to be omitted or cut or to be covered in a generic and less wholistic manner. The final issue is ***cultural awareness***. Professional military education often takes place in a setting where students from a wide variety of countries congregate and so ensuring that material is challenging whilst being aware of different social and political attitudes matters.

These five challenges relate especially when it comes to education on the importance of gender mainstreaming. Because students and staff are pressed for time, extra topics that may be 'nice to have' but are not seen as essential struggle to be included. In terms of interest, colleagues have their own areas of expertise and students tend to be particularly focused on war as a tool of state policy, rather than a social phenomenon with an array of consequences. When it comes to professional relevance, the army's role in kinetic engagements is widely accepted, but their role in protection of civilians and provision of assistance is less so. The institution, like all institutions, has its own biases and gender equality tends to generate discussions about meritocracy and whether women should serve in all army roles. As an academy where almost a fifth of students are from overseas, there is a need to be sensitive of different perspectives of the roles and vulnerabilities of women and girls. What then are the best ways of integrating the WPS agenda in military education while meeting these five challenges? The answer is to take an incremental and whole-of-course approach to education that effectively integrates the WPS agenda in all aspects of the course and highlights the implied role of WPS in responding to threats and challenges.

Integrating the WPS Agenda into PME – The Sandhurst Approach

In 2019, the department of Defence and International Affairs (DIA) sought to integrate the headlines of JSP1325 into the course, assessment and delivery. In particular, the core principles of UNSCR 1325 needed to be included and centralised. WPS was not something taught explicitly, and it was not integrated into academic or practical exercises in a meaningful manner. It was something DIA should have been doing, but discussions around teaching critical, and especially feminist approaches to international relations, and the WPS agenda prior to 2019 had been dismissed as non-essential in prior course updates by academic and military colleagues.

Integrating WPS education into PME was an incremental process and is still evolving. The following section highlights how this integration took place, by implementing what this chapter describes as the Sandhurst Approach.

Innovative Academic Exercise: Ex Complex Terrain

To start with the final summative assessment may appear counter-intuitive when it comes to education but integrating the WPS agenda into a final academic assessment has been an effective way of demonstrating its importance to students, military colleagues, and academic colleagues. It is worth noting that this approach bears many similarities with the 'backwards design model' developed by Grant Wiggins and Jay McTighe.[17] Although at this stage it wasn't the entire course that was being redesigned, it allowed for the establishment of new course aims (via the new assessment) which would then become extremely useful, as will be discussed later. In 2019, DIA replaced a 2000-word individual academic essay with a two-part assessment that involved a group Conflict Analysis Report (CAR) and an individual Human Security Individual Assessment (HSIA). These parts are equally weighted and constitute a critical component of the overall commissioning course at RMAS (students have to achieve a passing grade to commission into the Army). It concludes with a compulsory reflections class where students read one another's CARs and critique them as well as refining their own work. This did not happen in isolation, and it necessitated a small update to individual lessons in the course (see Table 9.2 below), but the assessment shift drove both the small course update and the inclusion of some WPS-centred readings.

The assessment ***Ex Complex Terrain*** enables OCdts to apply their understanding of security and conflict management to a complex and

17 For an overview of this approach see:
Indiana University Bloomington Center for Innovative Teaching and learning, "Backward Course Design. "https://citl.indiana.edu/teaching-resources/course-design/backward-course-design/index.html.
Grant Wiggins and Jay McTighe, *Understanding by Design* (Alexandria VA: Association for Supervision and Curriculum Development, 2nd Edition, 2005).

contemporary conflict and forces them to focus on the human terrain, which included the impact of the conflict on individuals and its differing effects on men, women, boys, and girls. Each intake assesses a different assigned conflict. Appendix 2 gives an overview of Ex Complex Terrain with the guidance and handout provided to students for Ex Complex Terrain: Ukraine, completed by students April-May 2022.

The best reports consider aspects such as CRSV, child soldiers, disruptions to health and education services, and organised crime, including modern-day slavery and human trafficking. The CAR requires students not only to conduct their own research, prioritizing primary documents from NGOs, CSOs, and IGOs, but also to develop their own approach and analytic framework to a contemporary conflict that enables identification of causes, drivers, actors, and the consequences of the conflict for different groups. To ensure professional relevance, students were directed to two types of core readings to be completed prior to the assessment. Firstly, they were instructed to read guidance published by the UK Stabilisation Unit to gain an understanding of the UK's approach to conflict analysis. Secondly, students were instructed to read the published doctrine on human security in military operations (JSP1325 then JSP985).

The submission of the CAR is followed by a timed and open book HSIA where students were asked to "Evaluate and Contrast the security implications for two metrics, X and Y" where X and Y were chosen in advance from a list of metrics that related to human security and the women, peace and security agenda. These were as follows:

1. Weapon proliferation and disarmament
2. Women and children
3. Refugees and Internally Displaced Persons
4. Economy and trade
5. State legitimacy
6. Violent extremism and terrorism
7. Organised crime
8. Environment
9. Health
10. Food, water, and other essential resources

This allows students to demonstrate an individual understanding of the conflict and means that some research on all ten aspects prior to the assessment would need to be undertaken. By centralising human security and within it WPS in a final summative assessment, it enables students to discover for themselves the impact how conflict relates to the WPS agenda. No specific training and education on the assigned conflict is offered, rather students apply concepts, theories, and approaches to conflict analysis themselves. The assessment is research led and encourages creativity through developing their own approaches to conflict analysis.

Ex Complex Terrain has been widely commended for its innovation and relevance to professional operations and has been taken up by other professional military academies as a manner of developing students analysis skills while getting them to do something that is professionally relevant. RMAS students have now assessed conflicts in Syria, Yemen, Myanmar, Mali, Nagorno-Karabakh, the Tigray province of Ethiopia, and Venezuela. The ninth and most recent case is Ukraine and the training enabled students to see beyond state-centred causes and implications of nuclear security and to look at violence against civilians, organised crime, refugees, and the gendered implications of the conflict. Ex Complex Terrain has become a major selling point of the course and invariably receives excellent feedback from staff and students.

This innovative assessment has been highly effective at overcoming the five key challenges to the inclusion of the WPS agenda presented earlier. In regard to ***time***, the inclusion of Ex Complex Terrain required minimal changes to the timetable (as will be illustrated later) and resulted in a slight reduction in marking workload for DIA colleagues when compared to the previous individual essay. In terms of ***interest***, Ex Complex Terrain case studies were selected based on which conflict the students were most engaged with and was most newsworthy. An entire intake of 180-270 students focuses on the same conflict each term and so this encouraged a wide and general understanding of the conflict and discussion outside of class. In terms of ***professional relevance*** – Ex Complex Terrain and its brief was centred on doctrine (JSP1325 then JSP985) and the UK Stabilisation Unit's approach to conflict analysis (via the Joint Analysis of Conflict and Stability framework) and insights about the WPS agenda were centred in their practical as well as their academic exercises. In terms of ***institutional biases***, the inclusion of this assessment contributed to the recognition of the importance of human security and WPS in military operation for both OCdts and colleagues, specifically thanks to the HSIA. In terms of ***cultural sensitivities***, great care and consideration was given when selecting the conflict to study in order to balance a desire for professional relevance with cultural sensitivities of the OCdts.

WPS on the Ground – Practical Exercises

At its core, the military is about 'doing' and relies heavily on training and standard operating procedures to keep high levels of operational effectiveness. Turning to the implementation of WPS-related consideration in operational planning, UK doctrine included clear framework and lists of questions that were designed to help commanders explore issues related to human security and the WPS agenda at all stages of their operational planning.[18] However, PME must go beyond the mandated *training* requirements set out in doctrine and must focus on *educating* on the importance of the WPS agenda in order to ensure that there is a lasting impact. Officers need to be able to explain to their troops *why* it is important to consider human security and WPS even in tactical field operations and, as such, a simple focus on training is not sufficient. To achieve this at RMAS, we have focused on integrating human security and the WPS agenda into practical as well as academic exercises. A summative practical exercise, Ex TEMPLER'S TRIUMPH, takes place in the middle of OCdt's final term (Week 36 of 44) and asks OCdts to plan and conduct stabilisation operations in a fictional conflict with a large 'civilian' population, this ties into the **prevention** pillar of WPS.

Ex TEMPLER'S TRIUMPH has 3 phases, each of which contains a set of 'serials' that challenge the OCdts to demonstrate leadership and make effective command decision on the ground: a **rural phase**, where OCdts operate in a rural area with a small civilian population; an **urban phase,** where OCdts operate in a built-up urban location with a large civilian population, and finally a **CIVPOP phase,** where the OCdts play the role of the civilian population in the urban area. Working closely with the military chief instructor team, we have included a human security dimension to the narrative, shifting it away from solely focusing on kinetic enemy-centric activities. In 2020, the overarching narrative of the exercise was rewritten to include a water-shortage issue creating divides and fault lines amongst the civilian population to highlight that not all deployments are 'warfighting' and that many have a strong human dimension. The role players come from two sources: RMAS OCdts during the CIVPOP phase who have assigned roles; and external army troops present to provide support to the exercise who provided the local population for the rural phase of the exercise. The main limitation with including WPS-related considerations

18 JSP 985 Volume 2 (guidance) remains outstanding at time of writing.

in the scenario here is directly linked to the limited presence of women in the military, leading the fictional civilian population on the exercise to be constituted on average of 90 per cent male. Nevertheless, we have been able to include the WPS agenda in all three phases of the exercise, as set out below:

Rural Phase

In the rural phase, the smaller local population is played by, mainly male, external Army troops. To mitigate the impact of this and to highlight the importance of the **protection** pillar of the WPS agenda, the scenario was updated to include human trafficking, which is used to explain the absence of many women in the rural areas. This notional trafficking is attributed to a fictional criminal network, which is active in the black-market economy of the urban area. Training activities are organised in ways that require the OCdts to demonstrate gender awareness. If they adequately identify special security needs of the female population and engage with them in a meaningful matter (with the use of female OCdts to gain access to the female-only tents for example), the OCdts will be 'rewarded' with some valuable information clearly showing that gender awareness is not only the right thing to do, but it is also required for operational effectiveness.

Urban Phase

During the urban operation phase, the OCdts interact with a large civilian population played by other OCdts. Here, the scenario ranges from political rivalries to humanitarian assistance. Throughout this phase, WPS consideration have been included in various ways: key roles have been assigned to women which requires the OCdts to take into consideration their unique needs when conducting stakeholder engagement. For example, in their planning, OCdts must take into consideration the special requirements of women and must ensure they have at least one female searcher on the gate whenever members of the civilian population are expected to arrive at their patrol base. The humanitarian dimension also requires OCdts to take into consideration the needs of the women and girls during the planning stages of operations delivering humanitarian assistance. For example, OCdts are tasked with facilitating a water distribution event. In both the planning and execution stages of this, OCdts are expected to take account for the security and access needs of mothers of infants, pregnant women, injured or elderly individual, etc. The aim here is to highlight the importance and relevance of the **relief and recovery** pillar of the WPS agenda.

CIVPOP Phase

The final phase of this exercise is where the OCdts take on the roles of the civilian population. Here, each OCdt is given a 'character' which a clear family, socio-economic background, clan/identity, and set objectives to accomplish during this phase. OCdts are given considerable freedom of operation and are directed to act in whatever way they think is appropriate considering their character's background and backstory. Beyond being a fun few days of role-playing, this CIVPOP phase of the exercise might be one of the most important ones because this will be the first time most OCdts experience being on the 'receiving' end of a patrol. This phase is of great importance as it allows OCdts to see how their actions are received and interpreted by local populations. With regards to the integration of the WPS agenda during this phase, the focus is put on the **participation** pillar. Female OCdts are assigned a proportion of the key roles as family heads and leaders to ensure the OCdts move away from the assumption that men will always be the family leader. Moreover, some women combatants are also included in the enemy forces (fictional terrorist or insurgent organisations) to break the presumption that women are passive in conflict, taking on ancillary roles, and highlight the importance of **gender mainstreaming** as part of conflict management.

Throughout all of this, we have been able to include a practical implementation of the WPS agenda in military operations whilst overcoming issues related to the fact that only about 10 per cent of the available civilian population are women. This enabled moving beyond the ***training*** in order to truly ***educate security*** by focusing on the understanding and analysing of the *gendered* consequences and implications of military operations. Looking back at the five key challenges, DIA's involvement in practical field exercises has a significant positive impact. Firstly, with regards to ***time***, no additional exercises needed to be added to the programme. Instead, the WPS agenda and human security were included at the core of a long-standing and well-established exercise which also contributes to overcoming the other challenges. In terms of ***interest***, the inclusion of WPS considerations contributes to the establishment of a more realistic and engaging human terrain and scenario which makes the exercise more stimulating and thought-provoking for the OCdts (and staff). Looking at ***professional relevance***, the inclusion of human security throughout the entire exercise clearly illustrated to all OCdts the impotence of the WPS agenda 'on the ground', outside of a classroom setting. Building on this, the wide-ranging inclusion of the WPS agenda and human security in this field exercise contributes to the tackling of ***institutional biases*** by showing to both the academy directing staff and the OCdts that human security and gender mainstreaming are important, relevant, and key to effective military operations. Finally, in terms of ***cultural sensitivities***, great care was taken as to not fall into the realm of humour and caricature when integrating WPS in an environment that includes such a limited number of women since this could easily have counterproductive consequences.

Integrating WPS – A Whole of Course Approach

The DIA update our part of the commissioning course biennially and in 2019, an update took place to accommodate the new assessment and ensure readings were up to date, in line with the backwords course design model. Rather than having specific classes on the WPS agenda, we deliberately took an incremental approach and weaved readings and learning points on WPS within our existing course structure. As mentioned above, the WPS agenda fit neatly with the four main themes that we cover in DIA.

There was no radical transformation of course content, but through selecting readings and amending themes, we were able to integrate training on the WPS agenda in each class. It was not a question of simply adding WPS to the course, but rather subtly integrating it into a pre-existing structure. A good example is our seminar on peace processes, peace-making, and mediation. This session allowed us to include one of the main strands in the WPS agenda: the vital importance of women's **participation** in all aspects of peace-making. Table 9.2 illustrates how our 90-minute class looked before and after the integration of WPS.

The Key Learning Point (KLPs) of the session remains, but there is inclusion of women in the additional themes. In addition to updating the KLPs, the pre-lesson reading lists was reviewed to include sources on women's participation in peacebuilding; for mediation we included documents such as the Council on Foreign Relations online resources on "Women's Participation in Peace Processes". Similar updates were centralised in other classes, especially those on the concept of security, causes

Course Pre-WPS update (2018)	Course Post-WPS update (2019)
DIA 018: Peace Processes: Peace-making and Mediation (Israel-Palestine) **Key Learning Points** The essential theme of the lesson is to assess peace processes as an international conflict management tool, with reference to peace-making and third-party mediation. Additional themes, some or all of which may be considered in the lesson at the discretion of the tutor to achieve the overall aim, include: • Identify conditions where peace processes may be necessary and assess the costs and benefits of third-party mediation or "honest broking" in managing, halting, and preventing conflicts. • Evaluate the strengths and weaknesses of specific mediators, including: The United Nations, the United States, the European Union, the Quartet, and the Arab League. • Identify and assess conditions accounting for the failure and success of peace negotiations, including *inter alia*, political will, financial assistance, popular support, key individuals, and recognition and cooperation. • Identify and evaluate the nature and key issues of three interrelated but separate conflicts: Palestinian-Israeli; wider Arab-Israeli; Israel and the Islamic world and assess key events in each conflict. • Critically evaluate the key historic milestones in the post-Cold War Israeli-Palestinian peace process including the Oslo accords, the Camp David negotiations, the Road Map, and the Annapolis Conference. • Analyse the most recent efforts at conflict resolution, including the Mitchell-led and Kerry-led talks and policy shifts since 2017. • Critically assess the continued appropriateness of long-standing solutions to the Arab Israeli peace process, including the one and two state solutions. • Evaluate the potential implications of a both a successful and unsuccessful outcome to the Arab Israeli conflict on regional and international peace and stability.	DIA 018: Peace Processes: Peace-making and Mediation (Israel-Palestine) **Key Learning Points** The essential theme of the lesson is to assess peace processes as an international conflict management tool, with reference to peace-making and third-party mediation. Additional themes, some or all of which may be considered in the lesson at the discretion of the tutor to achieve the overall aim, include: • Identify conditions where peace processes may be necessary and assess the costs and benefits of third-party mediation or "honest broking" in managing, halting, and preventing conflicts. • *Consider the implications of women's participation or exclusion from peace processes.* • Identify and evaluate the nature and key issues of three interrelated but separate conflicts: Palestinian-Israeli; wider Arab-Israeli; Israel and the Islamic world and assess key events in each conflict. • Evaluate the strengths and weaknesses of specific mediators, including: the United Nations, the United States, the European Union, the Quartet, and the Arab League. • Identify and assess conditions accounting for the failure and success of peace negotiations, including *inter alia*, political will, financial assistance, popular support, key individuals, and recognition and cooperation. • Critically assess the continued appropriateness of long-standing solutions to the Arab Israeli peace process, including the one and two state solutions. • Evaluate the potential implications of a both a successful and unsuccessful outcome to the Arab Israeli conflict on regional and international peace and stability.

Table 9.2. Comparison of Pre- and Post-WPS Update Lesson Overview

of conflict, democracy, and transnational security phenomena. We found this to be a particularly productive way of demonstrating the importance of the WPS agenda and considering the gendered impact of conflict without radically departing from the course framework that existed prior to the

	2017	2019	2021
1	Course Introduction and Conflict Analysis	Course Introduction and Conflict Analysis	Course Introduction and Conflict Analysis
2	Defining the International Security Environment	Defining the International Security Environment	The International System
3	Globalization, Nationalism, and International Order	Globalization, Nationalism, and International Order	Security 1: Definition and Developments
4	The Nature of Security	The Nature of Security	Security 2: Global Strategic Trends
5	Causes of Conflict	Causes of Conflict	Conflict 1: Theories of Conflict
6	Power and the Balance of Power	Power and the Balance of Power	Conflict 2: The Changing Character of Conflict
7	American Power and Hegemony	American Power and **the Liberal World Order**	New Frontiers of (in)Security
8	Rising and Resurgent Powers: Russia and China	Rising and Resurgent Powers: Russia and China	Transnational Security Phenomena
9	Fragile States and Intrastate Conflict	Fragile States and Intrastate Conflict	The Contemporary Terrorist Threat
10	Fragile States Case Studies	Fragile States Case Studies	Group presentation Support
11	The Nature of Terrorism	The Nature of Terrorism	Power and US Hegemony
12	The contemporary Terrorist Threat	The contemporary Terrorist Threat	Great Power Competition 1: Russia and China
13	Transnational Security Phenomena	Transnational Security Phenomena	Great Power Competition 2: Regional Competition and Strategic
14	Presentation Support	Presentation Support	Great Power Competition 3: State Interests and State Behaviour
15	Group Presentations	Group Presentations	Group Presentations
16	Stabilisation and Conflict Management	Stabilisation and Conflict Management	Conflict management and Ex Complex Terrain brief
17	The United Nations and Regional Organisations	**Democracy and Human Security**	State legitimacy, Democracy, and authority
18	Containing Nuclear proliferation and Deterrence	Containing Nuclear proliferation and Deterrence	Conflict Analysis Task
19	Peace Processes: Peace-making and Mediation	Peace Processes: Peace-making and Mediation	Human Security
20	International Law and the Use of Force	International Law and the Use of Force	Counterterrorism
21	Humanitarian Intervention and R2P	Humanitarian Intervention and R2P	Conflict Transformation Tools
22	Democratic Peace Theory and Democratisation	**Exercise Support**	Ex Complex Terrain: Support
23	Counterterrorism	Counterterrorism	Jus ad Bellum and R2P
24	Counterterrorism: CONTEST	Counterterrorism: CONTEST	Peacekeeping, Enforcement and Stabilisation
25	Alliances and Coalitions	**EX COMPLEX TERRAIN**	Contemporary Interventions
26	Strategic Thinking: Enablers and Constraints	Strategic Thinking: Enablers and Constraints	EX COMPLEX TERRAIN
27	The UK National Security Strategy and SRSD	The UK National Security Strategy and SRSD	UK National Security Strategies
28	UK foreign policy and key relationships	UK foreign policy and key relationships	UK Foreign and Defence Policy 1: Enablers and Capabilities
29	UK defence policy: key debates and challenges	UK defence policy: key debates and challenges	UK Foreign and Defence Policy 2: Constraints and Activities
30	Reflections: UK defence in the international security environment	Reflections: UK defence in the international security environment	Ex Complex Terrain and Reflections on Future Trends

Table 9.3. Comparison of DIA Course Content in 2017, 2019, and 2021

update. It also enabled colleagues who were less familiar with the WPS agenda and gender mainstreaming to consider each topic on its own.

A more substantive course update took place in 2021. Once the relevance and efficacy of Ex Complex Terrain had been established and feedback had been incorporated into the exercise to refine and improve it, it was the appropriate time to update the course in a manner that fed into the final summative exercise more explicitly and drew on the lessons learned during the pandemic and the opportunities for virtual learning. The course content covered by DIA are listed in three boxes – 2017 (pre-integration of WPS), 2019 (change of assessment), and 2021 (more wholistic focus on the WPS agenda) in Table 9.3.

As can be seen, the course changes from 2017 and 2019 (in bold) were limited to where it was vital for the success of Ex Complex Terrain. Sessions for exercise support and the written assessment necessitated the loss of two sessions and the material was incorporated elsewhere.[19] The 2021 update reflects the shift from JSP1325 to JSP895 and various other UK documents and strategies produced in the interim including the Integrated Review[20] and the latest update of Global Strategic Trends.[21] This update more effectively incorporates gender awareness and WPS in all aspects of the course. The security package from periods 1-10 incorporates the shifting concepts of security with a reinvigorated focus on WPS. Periods 16-25 feed more consistently into the summative exercise and include an individual conflict analysis task where the impact of a conflict on women, children and other vulnerable groups are considered individually. This course update may have met resistance in 2018, however, by gradually integrating WPS within the course, its relevance and importance has become truly embedded within the academy.

How then did the DIA's incremental approach meet the five challenges. In terms of ***time***, the assessment was designed to take no longer than a 2000-word essay but to encourage teamwork and a better

19 The session on United Nations and Regional Organisations was incorporated into a session on "American Power and the Liberal World Order"; Democratisation and DPT was brough up to session 17 to make space for Exercise Support; and the content of the session on "Alliances and Coalitions" was covered as part of assessing the Liberal World Order and UK foreign policy.

20 HM Government, "Global Britain in a Competitive Age: The Integrated Review of Security, Defence, Development and Foreign Policy", CP 403, 16 March 2021. https://assets.publishing.service.gov.uk/government/uploads/system/uploads/attachment_data/file/975077/Global_Britain_in_a_Competitive_Age-_the_Integrated_Review_of_Security__Defence__Development_and_Foreign_Policy.pdf.

21 UK Ministry of Defence, "Global Strategic Trends: The Future Starts Today".

use of OCdt time. No extra periods were added to the course and WPS was subtly integrated into the 2019 update. In terms of ***interest***, including WPS and human security through a broad range of lessons and tasks highlights the relevance of the topic and makes many of the issues being discussed in class more directly relevant to the lives and future careers of the OCdts. In terms of ***professional relevance***, embedding human security and WPS throughout the entire course significantly highlighted its direct professional relevance and mitigated any questioning or scepticism that might still persist. In terms of ***institutional biases***, changes were instituted incrementally and the pedagogic value of the assessment as an academic innovation was highlighted through termly qualitative feedback. In terms of ***cultural sensitivities***, there was a deliberate attempt to avoid cases where current students were involved, and seminars were used to explore these themes in a non-prescriptive manner.

Concluding Recommendations

Our experience in integrating WPS training within PME at RMAS may not work for every academic institution, but we believe that there are challenges that all PME establishments share. As such we set out these best practices for institutions seeking to integrate the WPS agenda into their education. The Sandhurst Approach draws on our own experience and makes 4 concrete recommendations:

1. **Start at the end**: identifying objectives is vital to any successful strategy. Being clear about what you want to achieve through integrating the WPS agenda in military education is vital for success. In academic settings, this may involve starting with the assessment and building up lectures and reading lists to that end. Devising an innovative assessment that saves time, encourages teamwork, engages students from various disciplines, has professional relevance, and meets the training objectives of your institution sets the foundation for the WPS agenda to be integrated in many parts of the training. Students and staff will spend a disproportionate amount of their study thinking about (and marking) the assessment and so making the assessment as interesting as possible is the best way of integrating WPS education into courses.

Challenge	WPS Training Lessons
Time	Ensure the WPS agenda is integrated throughout the course in ways that create more time for reflection. For example, WPS training is not an added burden for students and colleagues.
Interest	Ensure practical examples are used where gender mainstreaming has led to concrete outcomes that can be measured and related to military efficacy. Tie the WPS pillars to conflict analysis and choose up-to-date and relevant cases studies.
Professional Relevance	Integrate WPS scenarios and serials into practical as well as academic exercises. Ensure students think about gender mainstreaming in terms of operational effectiveness as well as gender equality.
Institutional Biases	Work with the institution and their priorities and don't be too academic about it – WPS education does not require a comprehensive course on feminist and critical approaches to IR. Go for an incremental approach.
Cultural Awareness	Emphasise the universal applicability of gender equality while simultaneously highlighting that gendered roles differ based on context. Ensure a variety of examples of how participation, prevention, and protection work in different contexts and, in the context of defence engagement, avoid too much emphasis on the failures and shortcomings of the country/region you are delivering in.

Table 9.4. PME Challenges and the Sandhurst Approach

2. **Don't 'Add WPS and Stir':** there is a long-established literature on the problems with militaries and politicians simply adding women and stirring to achieve numerical and quota objectives. However, we would argue that though it is vital to add WPS to PME courses, its addition should be targeted, incremental and smart. Students and staff are often unwilling to engage with new material and/or topics simply because of top-down directives and as such, targeting how, when, and why the WPS agenda is being added to each topic is a sounder way of integrating it into PME.
3. **Recognise your institutional limits:** we had an advantage in that the publication of JSP1325 opened space for integrating WPS without us having to justify its inclusion to our military colleagues. However, even here, DIA had few experts on WPS and no specialisms in the study of gender in international relations. An incremental approach is a sound way of bringing all stakeholders with you but is also sometimes a necessity. Starting slowly by simply referencing UNSCR 1325, including case studies

involving women's participation in peacebuilding, and asking students to consider the disparate impact of conflict on men and women, is a good place to start.

4. **Take every opportunity**: there is a tendency for military academies and defence setters to see Human Security and the WPS agenda as something to add to training. Where your institution gives you an inch, run with it.

The key recommendations of the Sandhurst Approach are summarised in Table 9.4, which can be used as a handrail when including WPS agenda in training courses.

The Sandhurst Approach that we have trailed and refined is in constant evolution, but it offers some thoughts on best practices for PME institutions that want to go beyond offering training and rather offer a full education in the Women, Peace and Security agenda.

References

Army, "RMA Sandhurst: Faculty for the Study of Leadership, Security, and Warfare", https://www.army.mod.uk/who-we-are/our-schools-and-colleges/rma-sandhurst/faculty-for-the-study-of-leadership-security-and-warfare/

Cooper, Erin. "Teaching the WPS Agenda in a Changing Security Landscape", Our Secure Future Policy Brief, January 2021. https://oursecurefuture.org/sites/default/files/Teaching-WPS-Agenda-Changing-Security-Landscape.pdf

Gender Action for Peace and Security UK. "Assessing UK Government Action on Women, Peace and Security in 2020", April 2021. https://gaps-uk.org/wp-content/uploads/2021/04/GAPS-Shadow-Report-2020.pdf

HM Government, "Global Britain in a Competitive Age: The Integrated Review of Security, Defence, Development and Foreign Policy", CP 403, 16 March 2021. https://assets.publishing.service.gov.uk/government/uploads/system/uploads/attachment_data/file/975077/Global_Britain_in_a_Competitive_Age-_the_Integrated_Review_of_Security__Defence__Development_and_Foreign_Policy.pdf

Indiana University Bloomington Center for Innovative Teaching and learning, "Backward Course Design. "https://citl.indiana.edu/teaching-resources/course-design/backward-course-design/index.html

International Committee of the Red Cross. Protocol Additional to the Geneva Conventions of 12 August 1949, and relating to the Protection of Victims of International Armed Conflicts (Protocol I), 8 June 1977. https://ihl-databases.icrc.org/ihl/INTRO/470.

Peace Women. "1325 National Action Plans (NAPs)". https://1325naps.peacewomen.org.

UK Ministry of Defence, "Global Strategic Trends: The Future Starts Today" 6th edition, 2 October 2018. https://assets.publishing.service.gov.uk/government/uploads/system/uploads/attachment_data/file/1075981/GST_the_future_starts_today.pdf

UK Ministry of Defence. "JSP 985: Human Security in Defence" December 2021. https://assets.publishing.service.gov.uk/government/uploads/system/uploads/attachment_data/file/1040257/20211209_JSP_985_Vol_1.pdf

UK Stabilisation Unit, "Joint Analysis of Conflict and Stability (JACS) Guidance Note", 1 June 2017. https://assets.publishing.service.gov.uk/government/uploads/system/uploads/attachment_data/file/765448/JACS_Guidance_Note.pdf

UK Stabilisation Unit, "The UK Government's approach to Stabilisation: A guide for policy makers and practitioners", 19 December 2018. https://assets.publishing.service.gov.uk/government/uploads/system/uploads/attachment_data/file/784001/The_UK_Government_s_Approach_to_Stabilisation_A_guide_for_policy_makers_and_practitioners.pdf

United Nations Security Council. *Security Council Resolution 1325*, 31 October 2000, S/RES/1325 (2000). http://unscr.com/en/resolutions/doc/1325.

United Nations Security Council. "Statement by the President of the Security Council", S/PRST/2002/32, 31 October 2002. https://www.securitycouncilreport.org/atf/cf/%7B65BFCF9B-6D27-4E9C-8CD3-CF6E4FF96FF9%7D/WPS%20SPRST200232.pdf

United Nations Security Council. "Statement by the President of the Security Council", S/PRST/2004/40, 28 October 2004. https://www.securitycouncilreport.org/atf/cf/%7B65BFCF9B-6D27-4E9C-8CD3-CF6E4FF96FF9%7D/Disarm%20SPRST200440.pdf

United Nations Security Council. "Statement by the President of the Security Council", S/PRST/2005/52, 27 October 2005. https://www.securitycouncilreport.org/atf/cf/%7B65BFCF9B-6D27-4E9C-8CD3-CF6E4FF96FF9%7D/WPS%20SPRST200552.pdf

Witter, Jennifer. "Mainstreaming WPS in the Armed Forces: The Case of Australia" in *The Oxford Handbook of Women, Peace, and Security*, Eds. Sara E. Davies and Jacqui True (Oxford: Oxford University Press, 2019), 10-11. https://jenniferwittwer.com/wp-content/uploads/2019/01/oxfordhb-WPS-in-Armed-Forces.pdf

Wiggins, Grant and Jay McTighe, *Understanding by Design*, 2nd Edition (Alexandria VA: Association for Supervision and Curriculum Development, 2005).

EPILOGUE

In September 2022, the Iranian morality police arrested and killed 22-year-old Jina Mahsa Amini, a young Iranian woman who dared to challenge the dress code enforced by the regime. The courage of the subsequent protestors is a global inspiration, and another reminder of the inequalities women continue to confront globally, as well as their vital role as active participants for change. Supporting women has never been more urgent. Their meaningful participation in peace has borne fruit and has been shown to be essential all over the world. Conflict-related sexual violence and domestic violence have been highlighted and centralized in conflict transformation activities and the inclusion of gender advisors within peace operations has gone some way to achieving the aims of the WPS agenda.

The UK is at the forefront of implementing the WPS agenda. It was one of its first signatories and has demonstrated enduring commitment through four National Action Plans. The UK also 'holds the pen' for WPS at the UN Security Council. In 2012, UK Foreign Secretary William Hague launched the Preventing Sexual Violence in Conflict Initiative with Angelina Jolie which led to over two thirds of all UN members endorsing the Declaration of Commitment to End Sexual Violence in Conflict in 2013. This was followed by an International Summit in London in 2014 attracting around 1,700 delegates from all over the world. Following on, in November 2022, the UK hosted an International Ministerial Conference on Preventing Sexual Violence in Conflict Initiative after which more than 50 countries and the UN agreed on urgent actions to end sexual violence in conflict. They signed up to a new declaration, which included national commitments and practical steps to address the triggers of gender inequality, the stigma faced by victims, and ensure that their national laws reflected the aims of WPS.

This book serves as a reminder of how far we have come, but also how much work is still to be done. The resurgence of war in Europe and preparing our servicepeople for the challenges this creates must not blind us to the gendered dimension of conflict and the human toll. In Ukraine, reports of conflict-related sexual violence against women and girls horrify, but also must strengthen our resolve to double down on understanding and responding to the disproportionate impacts of conflict on women and girls. As the annual report to Parliament on the UK National Action Plan on WPS stated in July:

> It is critical we keep the plight of Ukrainian women and children at the forefront of our agenda, working with every arm of government, with civil society and our international partners.

A key player in realizing WPS is the military. The MoD is making strides to ensure that the WPS agenda is central to its operations. This has involved looking inwards as well as looking outwards. The work of the 77 Brigade Outreach Group, where I am an honorary Colonel, is a crucial part of implementing the agenda and we are better training our people in gender awareness and mainstreaming. Fostering an understanding of how militaries can better implement the WPS agenda in crucial and it was my pleasure to attend and speak at the RMAS symposium. However, the WPS spans across government departments and requires a whole-of-government approach.

In 2023, the UK will produce its fifth National Action Plan, which will be owned by the MoD and FCDO jointly. It does so against the backdrop of a complex strategic environment where there are challenges but also opportunities for the UK to be a global leader on WPS. The 2020 Integrated Review calls for the UK to "shape the open international order of the future." It demands that the UK to be a force for good in the world and a major part of this is promoting gender equality. Our actions must meet this ambition. This requires a greater understanding of the moral urgency and the vital national interests served by ensuring the integration and implementation of WPS at home and abroad. The All Party Parliamentary Group on Women, Peace & Security (APPG on WPS), which I co-Chair with Flick Drummond MP, has been working in partnership to:

- Raise the profile of Women, Peace and Security issues across Westminster and Whitehall;

- Support and creates strong and vibrant networks to improve collaborative working amongst our members and partners;
- Share information, lessons learned and best practice;
- Strengthen the UK Government's implementation of, and approach to Women, Peace and Security;
- Build the understanding of Parliamentarians and civil society to support their effective advocacy, advice and scrutiny of relevant UK Government departments.

We have run a vibrant diverse programme of events, often partnering with other APPGs and NGOs. The APPG events have included looking at WPS in Afghanistan, Yemen, Iraq, the Philippines, women's participation in peace processes, preventing sexual violence in conflict, and we have also convened a roundtable of WPS leads in Embassies and High Commissions with similar outlook to the UK. Looking forward, the UK must learn the lessons from the past and devise new strategies for engaging with women's and civil society groups to realize the aspirations of the WPS agenda.

This book started with a challenge to "have the courage to reflect on the harsh lessons and failures of the Afghani experience…to heed the calls of women…to salvage the soul of democratic nations". The UK and our democratic allies must meet this challenge. I conclude, therefore, with a note of optimism. Just as the agenda has been enshrined in domestic legislation around the world, the institutionalisation of the agenda in UK law is underway. In June 2022, I introduced the Women, Peace and Security Bill in the House of Lords which, as I write, awaits its Committee Stage. If successful, this will embed WPS into UK law, thus ensuring it becomes an integral 'part of the DNA' in all FCDO and MoD activity undertaken by future UK governments regardless of party or personality. Peace and security can never be achieved if 50 per cent of the population are excluded and ignored and a renewed commitment to the WPS agenda is both a moral and strategic necessity.

Baroness (Fiona) Hodgson of Abinger CBE ,
Co-Chair of All Party Parliamentary Group on Women, Peace & Security
and Hon Col 77th Brigade Outreach Group

APPENDIX 1: NOTIONAL GENDER INEQUALITY OVERLAY CRITERIA INTERVALS

Taking the Gender Inequality overlay for the proposed model, this Appendix sets out notional intervals for the intensity of the prevalence of the different criteria that would be used to populate it.

Criterion 1.1, Average Educational Deficit: Average Years of Schooling (AYS) for Women Measured at Age 25 (300 Months)	
Class 1 (Highest Risk)	0–2.99 AYS
Class 2	3.0–5.99 AYS
Class 3	6.0–8.99 AYS
Class 4	9.0–11.99 AYS
Class 5 (Lowest Risk)	>12.0 AYS
Criterion 1.2, Labour Force Participation: The Ratio of Percentage of Female and Male Labour Force Participants Between Ages 14–65 (168-780 Months)	
Class 1 (Highest Risk)	0–10% Female/90–100% Male
Class 2	10–20% Female/80–90% Male
Class 3	30–40% Female/70–80% Male
Class 4	40–50% Female/50–60% Male
Class 5 (Lowest Risk)	50% or more Female/50% or less Male

Criterion 1.3, Maternal Mortality Ratio: Number of Maternal Deaths for Every 100,000 Live Births	
Class 1 (Highest Risk)	>900.0 Maternal Deaths/100,000 Live Births
Class 2	600.0–899.99 Maternal Deaths/100,000 Live Births
Class 3	300.0–599.99 Maternal Deaths/100,000 Live Births
Class 4	150.0–299.99 Maternal Deaths/100,000 Live Births
Class 5	25.0–149.99 Maternal Deaths/100,000 Live Births
Class 6 (Lowest Risk)	<25.0 Maternal Deaths/100,000 Live Births

Criterion 1.4, Gender Differentiated Literacy Rates: Per Cent Difference of Literacy Between Males and Females Age 15 and Older	
Class 1 (Highest Risk)	>45.0% Difference
Class 2	35.0–44.99% Difference
Class 3	25.0–34.99% Difference
Class 4	12.0–24.99% Difference
Class 5	0.01–11.99% Difference
Class 6 (Lowest Risk)	0.0% Difference

Criterion 1.5, Gendered Cultural Violence	
Class 1 (Highest Risk)	>50% of Women Charged Plead Self-Defence (SD)
Class 2	40.0–49.99% of Women Charged Plead SD
Class 3	30.0–39.99% of Women Charged Plead SD
Class 4	20.0–29.99% of Women Charged Plead SD
Class 5	10.0–19.99% of Women Charged Plead SD
Class 6 (Lowest Risk)	0.0–9.99% of Women Charged Plead SD

APPENDIX 2: EX COMPLEX TERRAIN

Title: Ex COMPLEX TERRAIN: Ukraine

Aims: The aim of **Ex COMPLEX TERRAIN: Ukraine** is to assess the student's ability to analyse and understand a complex conflict, drawing on the concepts covered in module 3, *The Nature of the International Security Environment* and the stabilisation tools covered in module 4, *International Security Responses*. Students are also required to identify key security implications of specific dimensions of conflict from a military perspective and prepare short briefs that can be actionable. Thus, this assessment will allow the students to put into application the knowledge accumulated in DIA, drawing links between the different seminars and their future careers.

This Exercise requires groups of students to organise their own work in their own time and is designed to evaluate whether the student has successfully achieved the following Intended Learning Outcomes.

Intended Learning Outcomes:

1. Critically evaluate the nature and applicability of peace and stability as concepts in contemporary international politics.
2. Assess a range of international responses; including peace processes, international negotiation, sanctions and the use of force in its many forms.
3. Critically evaluate UK foreign and defence policy as a contribution to stabilising the international system.
4. Construct and debate arguments and formulate original conclusions.

5. Plan work and communicate findings clearly, accurately and fluently in both written and oral formats
6. Reflect upon, offer assessment of and subsequently refine independent work and the work of others

Description of task(s) to be completed:

There are three components to the assessment of **Ex COMPLEX TERRAIN**

1. **Conflict Analysis Report** (CAR) (50% of final mark)
   In groups of 3 or 4, students will be asked to complete a Conflict Analysis report (CAR) for a specific conflict situation.
   1. Each group is required to submit a written report of their analysis of 2000 words excluding references, footnotes, bibliography, and visual aid(s).
   2. The conflict analysis **must** include a **visual illustration** of the dynamics of conflict ('mapping the conflict').
   3. The analysis needs to be supported by appropriate sources.
   4. **A full bibliography is required (lecturers will comment on a draft bibliography in Week 1 of Seniors term).**
2. **Individual Human Security Impact Assessment (HSIA)** (50% of final mark)
   Students will be asked to write an individual assessment in timed conditions (15 minutes of planning and one hour of writing) that evaluates and analyses the importance of two human security factors in the conflict they have analysed.
   1. Assessment question: *Evaluate and contrast the security implications of this conflict for X and Y.*
   2. 'X' and 'Y' will be selected from the list below:
      1. Weapon proliferation and disarmament
      2. Women and children
      3. Refugees and Internally Displaced Persons
      4. Economy and trade
      5. State legitimacy
      6. Violent extremism and terrorism
      7. Organised crime
      8. Environment
      9. Health
      10. Food, water, and other essential resources

The assessment is open book. It is advisable students have access to a copy of their group CAR and a one-page A4 "cheat sheet" of key information to assist them with key facts to support their conclusions. The assessment will be typed on personal computers and there is no set wordcount.

3. **Read and Review other reports for class discussion** (DIA029) Students will be required to read the other CARs in their syndicates to enable a discussion and refinement of their own work and the work of others. This does not affect the overall mark awarded; it is, however, a requirement for successful completion of DIA module.

Deliverables required and submission information:

1. A written CAR framework of 2000-words (excluding references, footnotes, bibliography, and visual representation of the conflict dynamics) containing critical analysis and an evidence-based argument. Work should be clearly written, structured logically and supported with appropriate wider research. Ensure that a visual representation of the conflict and a bibliography are provided, and the work is consistently referenced throughout. This **must** be emailed to syndicate tutor by Wednesday, Week 2, Senior Term.

Failure to provide a bibliography, clear referencing, providing a visual representation that does not demonstrate conflict dynamics adequately, or writing +/-10% of the word count will be subject to an automatic penalty of 5 marks for each violation.

Plagiarism or failure to provide a visual representation will result in a mark of 0 and will require an automatic resubmission in Week 12.

If a group provides a visual representation that is taken from elsewhere but is referenced, the group should receive a grade capped at 10 for "structure and delivery."

2. HSIA to take place in DIA024 on personal computers, open book with advice to use group report and one-page "cheat sheet" of key information.

Estimated time to complete:

It is the responsibility of the student to allocate an appropriate amount of time to this piece of work based on the duration of the module, the weighting of the coursework and the hand in date.

Required Readings:

- *Joint Analysis of Conflict and Stability: Guidance Note, Stabilisation Unit, June 2017.*
- *Analysis for Conflict and Stabilisation Interventions: What Works Series, Stabilisation Unit, October 2014.*
- *JSP985 Human Security in Defence Volume 1: Incorporating Human Security in the way we Operate*

No formal case specific readings will be provided. It is strongly recommended that students prioritise primary documents including: think tank reports; parliamentary and other government resources; in-depth newspaper analyses; and date amassed by relevant NGOs and IGOs.

ANNEX 1: JSP985: HUMAN SECURITY IN DEFENCE[1]

WPS is a key cross cutting theme for the UK within the wider context of Human Security. In December 2021 a new Joint Services Publication (JSP) was produced by the Ministry of Defence (MoD). **JSP985: Human Security in Defence** articulates how the MoD should incorporate a range of Human Security (HS) considerations (Factors & Cross Cutting Themes) – into training, planning, and military operations. These 8 HS Factors and a number of Cross-Cutting Themes (which align closely with NATO doctrine) are a basis for conflict analysis and effective military operations – these are illustrated in Table A-1:

Human Security Factors provide context to understand the human terrain / environment:

1. **Personal/Physical security** refers to individuals' and communities' freedom from actual or threat of physical violence from the state or external states, from violent individuals and sub-state actors, domestic violence, or other malign actors.
2. **Political security** refers to people living in a societal system that honours their basic human rights. Along with repressing individuals and groups, governments may try to exercise control over ideas, information, and freedom of expression.
3. **Economic security** refers to individuals' quality of life now and their future prospects related to their economic and financial circumstances. It links to the GDP of a state, job opportunities and education. It is a key driver of migration.

1 Annex prepared by: Lt Colonel David Kemmis-Betty MBE DL SCOTS, SO1 Human Security/Civ-Mil Interaction.

Integrate a range of human security considerations (factors & cross-cutting themes) into the way we understand, plan & operate / fight							
Freedom from fear – Freedom from want – Freedom from indignity							
Understand					Plan & Operate / Fight			Endstate & Outcomes
Areas	Structures	Capabilities	Organisations	People	Events			
PMESII	Human Security Considerations: HS 'Factors' (linked to UN SDGs)	HS Cross-Cutting 'Themes'	Conflict Dynamics	Plan Coord Integrate	HS Objectives	HS Tasks	
	Personal / Physical	Protection of civilians	Threats	J2 J3 JS Plans	Act as a 'force for good'	Engage	Heighten Understand
		Women peace & security (incl conflict related sexual violence & SEA)		Targets		-	
		Modern slavery & human trafficking	Vulnerabilities	Joint effects J7 Training	-		Improved integrated planning & targeting
		Preventing & countering violent extremism	Needs	J9 POLAD	Mitigate harm	RECOGNISE	
		Children affected by armed conflict / Youth peace & security	Griefances	LEGAD CIMIC HS Advisor MEDIA / IA		RESPOND (Caveat Ability to act depends on scope of mission, mandate & resources)	
P – Political	Political	Building integrity / Countering corruption			-		Improve fusion & engagement
M – Military			Motivations	sS / Joint			
E – Economic	Economic				Facilitate access to basic needs (FABN)		
S – Social / Cultural	Culture / Community	Cultural property protection	Narratives	X-Govt		REPORT (Up MIL CofC)	Conflict sensitive outcomes & reduced human rights violations / CIVCAS
I – Info*	Information*		Community dynamics	Partners Allies	-		
I – Infra*	Health			Host nation IOs			
	Food		Conflict drivers & consequences	NGOs CSOs	Contribute to safe & secure environment	REFER (on to HN, IOs, NGOs, CSO)s	Maintain legitimacy
	Environment / Climate						
Sex & age disaggregated data (SADD) analysis (How conflict impacts men, women, boys & girls / elderly) = gendered analysis							

Table A-1. Human Security (HS) – Concept Slide (Reference: JSP 985)

4. **Food security** refers to the physical and economic access to food and clean water. It is closely linked to agriculture and climate change and is a major cause of migration.
5. **Health security** refers to illness, disease and issues such as access to safe water, health services and gender specific services, and provision of basic knowledge to live a healthy life.
6. **Environmental/Climate security** refers to threats that environmental events and trends pose to individuals, communities, or states. Examples include deforestation, pollution, forest fires, drought or heat waves, access to water, floods, glacial melt, salination, polluted water and resource nationalism.
7. **Cultural/Community security** refers to a state in which individuals' traditional relationships and values alongside their tangible and intangible cultural heritage are not threatened by states or sectarian and ethnic violence with their traditions, values, freedom of religion, places of worship free from threat.
8. **Information security** refers to access to information, which both empowers (educates and provides opportunities) and can be manipulated (through extremist recruiting or dis-information, mis-information and malign-information). Information security also identifies the vulnerability to modern communication systems from cyber threats and information domination by single states or corporations.

Human Security Cross-Cutting Themes build on these HS factors, but focus more on violations of International Law and Human Rights:

- **Protection of Civilians** in conflict is central and links to LOAC obligations to Minimise Harm and where possible to Facilitate Access to Basic Needs and Contribute to a Safe and Secure Environment.
- **Women, Peace and Security**, refers to UNSCR 1325 and the UK National Action Plan on WPS. A key focus for the UK is Preventing Sexual Violence Initiative (PSVI) and particularly preventing Conflict Related Sexual Violence (CRSV), sex trafficking and a zero tolerance towards any form of Sexual Exploitation and Abuse (SEA) by security forces (including our Partner Forces) or humanitarian workers. WPS issues, cut across all other HS considerations.

- **Children Affected by Armed Conflict (& Youth Peace & Security),** is based on the rights of a child and preventing the 6 'grave violations': the killing and maiming of children, recruitment or use of children as soldiers, sexual violence against children, abduction of children, attacks against schools or hospitals, and denial of humanitarian access for children.
- **Modern Slavery and Human Trafficking** refers to the criminal networks and armed groups who take advantage of conflict to exploit vulnerable individuals to sell them for sex or as slaves, soldiers, forced labour or forced marriage. The military (G4) logistic chain needs to a take care when planning supply chains overseas to avoid inadvertently using contractors or sub-contractors who are directly or indirectly involved in slavery or are victims of slavery and trafficking in accordance with Modern Slavery Act (2015).
- **Building Integrity** is about countering corruption in conflict affected countries. Corruption feeds organised crime and fuels conflict.
- **Preventing and Countering Violent Extremism** is linked to the UK counter terrorism strategy (CONTEST) and seeks non-coercive methods to counter radical narratives and undermine pathways to extremism.
- **Cultural Property Protection** refers to our responsibility under the Hague Convention to protect both the tangible, visible cultural sites and expression of a community and the intangible such as the written word, stories, traditions & beliefs.

These factors and cross cutting themes are both drivers of conflict and consequences of conflict. Analysis of these considerations reflects a greater focus on the human terrain. Analysis should (where appropriate) be underpinned by **sex and age disaggregated data** to inform a 'gendered analysis'. Deeper understanding should be integrated into planning and enable UK Forces to use the '4R's to:

- **Recognise** HS related problems,
- **Respond** (where we can),
- **Report** issues up the chain of command and,
- **Refer** them on to responsible authorities (such as the host nation, UN or other IOs, NGOs or CSOs acting in the area).

In summary, Human Security planning considerations should be integrated across all staff functions, with advice from the nearest qualified Human Security Advisor or through reach-back to the Outreach Group (currently part of 77 Brigade but due to move to the 11th Security Force Assistance Brigade in 2023). There is a responsibility on all UK military personnel; soldiers, commanders, and staff officers to incorporate Human Security – as detailed in JSP 985 – into the way we train, understand, plan and operate or fight, to enable units to **Recognise** and **Respond** (where possible) and then **Report** and **Refer** Human Security related issues they encounter on the ground to the appropriate authorities. This will not only improve understanding, planning, targeting and fusion, but will also lead to improved conflict-sensitive outcomes, reduced human rights violations, and increased military effectiveness.

INDEX

Note: Numbers separated by commas indicate that the topic was mentioned in distinct ways and therefor warrants being listed in the index independently (rather than separated by a hyphen)

We hope that you have enjoyed this book.
Please consider posting a review for future readers.

Thank you.